GOOD VIBRATIONS

GOOD VIBRATIONS

The Physics of Music

Barry Parker

The Johns Hopkins University Press Baltimore

Printed in the United States of America on acid-free paper
2 4 6 8 9 7 5 3

The Johns Hopkins University Press
2715 North Charles Street
Baltimore, Maryland 21218-4363
www.press.jhu.edu

Library of Congress Cataloging-in-Publication Data

Parker, Barry R.
Good vibrations : the physics of music / Barry Parker.
p. cm.
Includes bibliographical references and index.
ISBN-13: 978-0-8018-9264-6 (hardcover : alk. paper)
ISBN-10: 0-8018-9264-3 (hardcover : alk. paper)
1. Vibration. 2. Sound-waves. 3. Sound.
4. Wave-motion, Theory of. I. Title.
QC231.P37 2009
781.2—dc22 2008054589

A catalog record for this book is available from the British Library.

The Johns Hopkins University Press uses environmentally friendly book materials, including recycled text paper that is composed of at least 30 percent post-consumer waste, whenever possible. All of our book papers are acid-free, and our jackets and covers are printed on paper with recycled content.

CONTENTS

ACKNOWLEDGMENTS

I am grateful to Trevor Lipscombe for his many suggestions and help in preparing this volume. I would like to thank Carolyn Moser for careful editing of the manuscript and the staff of the Johns Hopkins University Press for their assistance in bringing this project to completion. Thanks, too, to my artist, Lori Beer, for doing an excellent job on the drawings. Finally, I appreciate the assistance I received from Mike's Music of Pocatello in obtaining the photographs.

For more information on the physics of music and information on other books by the author, visit the web page www.BarryParkerbooks.com.

GOOD VIBRATIONS

Introduction

Physics and music may seem light years apart to many people. But surprisingly, they are closely related. Of course, music is sound, and sound is a branch of physics, but they are also connected in another way. Both are highly creative endeavors. The major advances in physics had their origin in somebody's mind. Einstein gave us relativity theory; Heisenberg and Schrödinger gave us quantum theory. In the same way, Beethoven gave us several magnificent symphonies, and Chopin gave us an array of beautiful piano pieces. So physics and music are both products of the mind. Physics may conjure up an image of difficult and complicated mathematics for some people, but to many it is a delightful and enjoyable endeavor. And certainly, most people love music. So I guess we can say that both have devoted fans.

This brings us to the question of whom this book is written for. I believe it will be of interest to musicians who are interested in learning more about the science behind music and to students and fans of physics, most of whom are also music lovers. (I know: I'm one of them.) But in writing this book, I had to consider one issue: there is a big range of tastes in music. Some people love classical music and hate rock; others love rock and hate classical. Because of this I've tried to steer a middle road. I talk about all types of music; in fact, I have a chapter that surveys all the music types. And I'm afraid that's about all I can do. I hope it satisfies most people.

A Short Definition of Music

Okay, let's begin by considering the question, What is music? It may seem like an odd question; after all, everyone knows what music is. But there is actually more to it than you may think; besides, it's useful to consider the question in that it may help you understand the relationship between music and physics more thoroughly.

We'll start with descriptions of music given in the dictionary: music is "the art of combining sounds with a view to beauty of form and the expression of emotion" and "the art of organizing tones to produce a coherent sequence of sounds intended to elicit an aesthetic response in a listener." Both of these descriptions give us a pretty good idea of what music is, but neither really goes to the heart of the question. Music is, in fact, very difficult to completely and satisfactorily describe. We know it is an art and that it is something that initiates an emotional response from almost anyone who listens to it. Beautiful passages can give you goose bumps, and terrible ones can illicit disgust or even rage. Regardless of your age, in fact, music elicits a response of some type in almost everyone. Even babies respond to music: a crying baby will quickly calm down when its mother begins to sing to it. Furthermore, it seems that the human brain is programmed to respond to music. Study after study has shown that there is a correlation between music (particularly music making) and the deepest workings of the brain. We now know that it is the right side of the brain that gives us our appreciation of music, but research has shown that music actually engages many areas of the brain. We know, for example, that music is important in relation to language development, and interestingly, it also appears to enhance mathematical ability. So besides its sheer entertainment value it may have other dividends.

You might ask, What characterizes music? As it turns out there are four major things. First of all, music is made up of tones, and these tones have a specific frequency, or *pitch;* in other words, some notes are high and some are low. Secondly, music has *rhythm,* which is related to the length of each tone. Basically, rhythm is the beat of the music, and this beat is usually regular (it's what you tap your feet to). Third, music varies in *intensity;* it can be loud or quiet or somewhere in between, but in almost all cases it varies to some degree. Finally, music has a quality that we discuss in some detail later in the book. It is called *timbre;* in a sense it's the thing that makes music interesting, but the easiest way to describe it is as follows: timbre is the property that allows us to distinguish various musical instruments even when somebody is playing the same tune on them. It's also what makes one voice different from another.

Another interesting question is, Where did music begin? Music is so basic that there's little doubt that even the earliest humans had some sort of music, or at least recognized rhythms. They may have banged on crude drums, or chanted, but this was a form of music. The earliest piece of written music was discovered in Mesopotamia (today's Iraq) and is dated about 1500 BC.

Pythagoras and Musical Scales

For music as we know it, we need a sequence of tones that are "linked"—in other words, tones that have a pleasing relationship with one another. We refer to a linked sequence of tones as a *scale.* From a simple point of view, we can say that a scale is a special set of notes that seem to belong together.

The first scale was developed in Croton in southern Italy. About 539 BC the Greek philosopher Pythagoras founded a school in Croton. Most people know Pythagoras for his famous formula relating the sides of a triangle, but he actually accomplished many other things. He had an intense interest in musical tones, but his major interest was not their beauty: it was how they related to numbers, and mathematics. According to a story that has been told many times, he was passing a blacksmith's shop one day when he heard the ringing of the blacksmith's hammer as it struck the anvil. He stopped to listen and soon began to wonder if the tone would change if the blacksmith applied more force. He therefore stepped into the shop and asked the blacksmith if he would hit the anvil with different forces. To his surprise, the force of the blow didn't change the tone; the only thing that had an effect was the weight of the hammer. Confused, he went home and looked into the issue further. He built himself a device we now refer to as a *monochord* (fig. 1). It was a hollow box with a string stretched across it; the string was supported by two wedges near the ends. Pythagoras presumably attached masses of different weight to the ends of the strings, but for simplicity we'll assume that

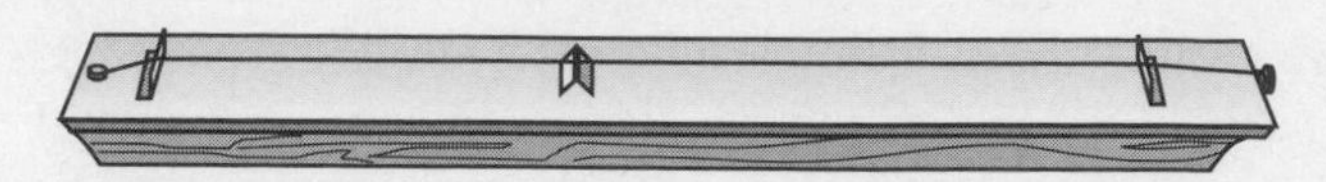

Fig. 1. Pythagoras's monochord.

the string was fixed at one end and had a screw at the other end that could be used to tighten it. When he plucked the string a certain tone sounded, and when he tightened the string, the tone changed. We know that when a string is tightened, the tone increases in pitch, but Pythagoras knew nothing about pitch. Nevertheless, he also discovered that a shorter string at the same tension would also change in pitch (the pitch would increase).

He then proceeded to do several experiments. His main concern was ratio of the tones, so he began comparing various tones. You can do this by sounding two tones together or one immediately after the other. His first experiment consisted of producing a tone from the entire unstopped string and comparing it to a tone produced when a stop was placed halfway along the string, as shown in figure 2. He found that the combination of tones was pleasing. The ratio was 2/1 (the original length divided by the length of half of the string). The *interval* between the two tones in this case is an octave, and as we'll see later, the octave is a basic unit of music. The notes (or tones) on the piano, for example, are arranged in octaves so that a given note in one octave is the same as a particular note in the octave above it (but at a different pitch).

In his second experiment Pythagoras stopped the string one-third of the way from one end (fig. 3), and compared the tones from it to one from the entire string (he plucked the section that was 2/3 of the length of the entire string). Again, he found that the two tones were pleasing when sounded together. We know that they are a *perfect fifth* apart; this is like striking middle C on the piano, then striking the G above it immediately after (or playing them together). When he plucked the shorter section (1/3 of the original length), it

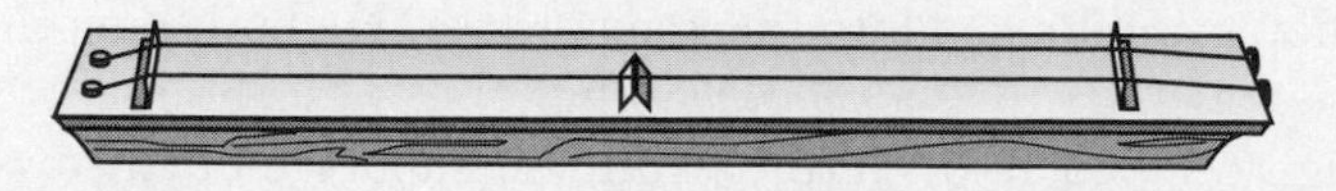

Fig. 2. The monochord with a stop halfway along the string. The full, unstopped string is also shown.

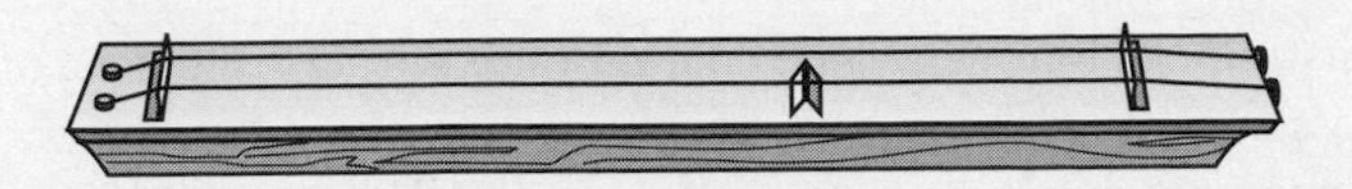

Fig. 3. The monochord with a stop one-third the way along the string.

also produced a pleasing tone with both the original length and 2/3 of it. It is G an octave above the lower G on the piano. As we will see, the perfect fifth plays an extremely important role in music.

In his next experiment Pythagoras compared the notes given by the 2/3 L string (L being the length of the original string) to the one from the 1/2 L string and noticed that they were also harmonious when sounded together. We know this as a *perfect fourth*. On the piano it is the interval between middle C and the F above it. What Pythagoras deduced from all this was that the integers 1, 2, 3, and 4 played an important role in making harmonious tones. He may have also noticed that the interval from C to E, which has a ratio of 5/4, was also harmonious, so we can extend this to the integers from 1 to 5.

Let's look more closely at the ratios he discovered, beginning with 1, and writing them as 1/1, 5/4, 4/3, 3/2, 2/1. Their decimal equivalent are 1.000, 1.250, 1.333, 1.500, 2.000. They represent four important musical intervals: a major third, a perfect fourth, a perfect fifth, and an octave. From them Pythagoras was able to devise a musical scale called the *pentatonic scale*. In chapter 6 we will look at how scales are set up in much more detail; in particular, we'll see how the pentatonic scale can be used to devise the eight-note scale we use today.

A Map of This Book

We turn now to a brief summary of what the book is about. Music is, of course, sound, and sound, in turn, is a wave, so the early part of the book is concerned with sound and its relation to waves. In chapter 1 we will see how sound is related to wave motion and how wave motion, in turn, is related to another phenomenon called simple harmonic motion. We will see that there are two basic types of waves and how sound is produced by one of them. Also, as I mentioned earlier, one of the major properties of sound is loudness or intensity, and I introduce a table of loudness called the decibel scale.

In chapter 2 we look at how we hear music. The parts of the ear are introduced, and there is a discussion of how we recognize pitch and distinguish loudness. The chapter ends with a discussion of hearing loss.

Chapter 3 discusses waves in more detail. In particular, we consider what happens when waves strike various types of boundaries. Two of the phenomena that occur are reflection and refraction; we will see how they are important in relation to sound. Another phenomenon of importance is interference, and we will also examine it. Finally, we will look at a type of wave that plays a central role in music, the standing wave.

In chapter 4 we get to music itself. In this chapter we apply what we have learned in the first few chapters to music. In particular, we talk about overtones and the timbre of music. Various vibrational modes will be considered, and we will look at what is called harmonic analysis.

In this introduction I have already introduced the scale, which is, of course, central to music. We return to this subject in chapter 5, where I discuss several scales in detail—the Pythagorean scale, the diatonic scale, the tempered scale, and major and minor scales, as well as two scales of particular interest to musicians today, the pentatonic and the blues scales.

Closely related to scales are chords and chord sequences, introduced in chapter 6. Here, I discuss in detail the many different kinds of chords and show you how to fill in a melody using chords, an important part of any musician's skills. We also look at other topics, such as chord sequences and the circle of fifths.

In chapter 7 we turn to rhythm and a survey of most of the types of music, ranging from rock and roll to the blues, jazz, new age, pop, and classical music. I think this chapter will give you a good idea of the large range of musical types.

Musical instruments are, of course, central to music, and in chapter 8 we begin a survey of them and the physics on which each is based. In this chapter we discuss the piano, tracing it from its beginnings; in particular, we look at the important contributions of Christophori. We also consider the construction of a piano, the role that the various strings play, and finally how a piano is tuned.

Another important stringed instrument is the violin. In chapter 9 we talk about both the violin and the guitar, along with some of the other stringed instruments. The art of violin making is important, and I discuss it along with the most famous of violins: the Stradivarius. We also look at the basic physics of violins and at some of the violin virtuosos. The chapter ends with a discussion of the guitar, which (as you no doubt know) is probably the most popular instrument in America today.

Chapters 10 and 11 cover the brass instruments—in particular, the trumpet and the trombone—and the woodwinds (particularly the clarinet and the saxophone). We look at the basic physics of each, how each instrument works, and as an aside, some of the outstanding instrumentalists.

One of the most important musical instruments is one we usually do not consider to be an instrument—the human voice. It is central to most types of music. In chapter 12 I begin with a history of singing, then go on to look at the parts of our anatomy that produce the singing voice, namely, the lungs and the vocal cords. Other topics considered in this chapter are phonetics, resonators, and the singing formant (a region of resonance). Finally, I also talk about some of the more famous singers of the last few decades.

Within the last few years (or perhaps I should say decades) music has changed significantly, and the major reason is the introduction of electronic instruments, particularly synthesizers. In chapter 13 we discuss electronic music. We look at the synthesizer and talk about the introduction of digital technology in music; this technology has, as you probably know, caused a revolution. I also introduce MIDI (Musical Instrument Digital Interface), which enables different electronic instruments to communicate with one another and has also changed music significantly. The chapter ends with a discussion of the microphone and loudspeakers, both of which are central to modern music.

MIDI plays such an important role in music today that it's worth looking at in detail, and we do this in chapter 14. It centers on the use of sequencers to record music. I will concentrate on the software approach, which is widely used today. The electronic recording industry has exploded in the last few years, particularly with the introduction of samplers, samples, and virtual instruments; they have, in fact, changed music recording significantly and are now used by literally everyone in the industry. I also discuss mixing, central to sequencing and recording, and the various sound effects such as reverberation.

In chapter 15 we turn to acoustics, both of concert halls and smaller studios. As we will see, Wallace Sabine of Harvard almost single-handedly developed the science around the turn of the century. Central to the acoustics of a concert hall is what is called the reverberation time (the time to inaudibility). We will look at the significance of reverberation in relation to concert halls, and I will show you how to calculate it. Also in this chapter we will consider what are usually referred to as "home studios," the smaller studios set up by amateurs or professionals that are playing an increasing role in the music industry.

In the epilogue I discuss iPods briefly because they are now playing a large role in the music industry.

Musical Notation

Throughout the book I use musical notation. Since not everyone may be familiar with it, the following is a brief survey of it. If you're already familiar with it you can, of course, skip this section.

In chapter 5, I discuss the musical scale in detail. Here, I give only a brief overview. In doing so, it is convenient to refer to the piano, but of course, everything I say also applies to other instruments. On the piano we have the notes as shown in figure 4. The low C here is usually referred to as middle C because it is roughly in the middle of the piano keyboard. The black keys are sharps and flats. For a given note—say G—the black key to the left of it is G-flat (written as G♭), and the black key to its right is G-sharp (written as G♯). (More generally, a sharp is a half tone up from a given note, and a flat, a half tone down, so that in actuality any key on the piano—black or white—can be the sharp or flat of an adjacent key.)

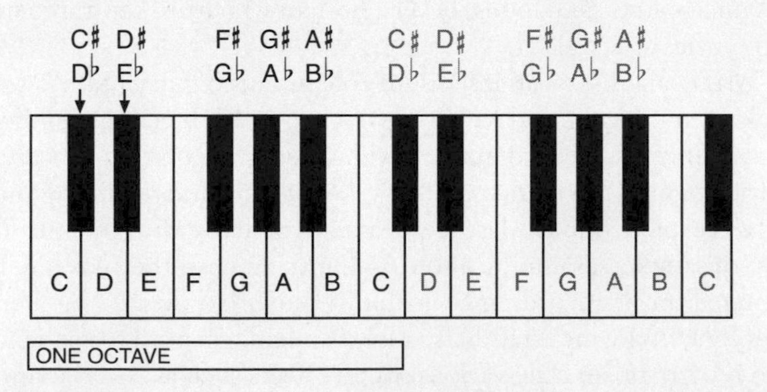

Fig. 4. The piano keyboard, showing the notes up from middle C. Two octaves are shown.

To designate these notes we have what is called a *score*, namely, the written music. The score is composed of sets of five horizontal lines, each of which is called a *stave*. Each of the lines on a stave, and the blank spaces between them, refer to notes on the piano (or any instrument). There are, in fact, two staves, one for the right hand (treble) and one for the left (bass). For the treble we have (where the sign 𝄞 is a treble clef)

This example shows what is called a whole note, but in practice many different types of notes appear. (I will discuss them below.)

For the left hand, or bass, we have (where 𝄢 is called the bass clef)

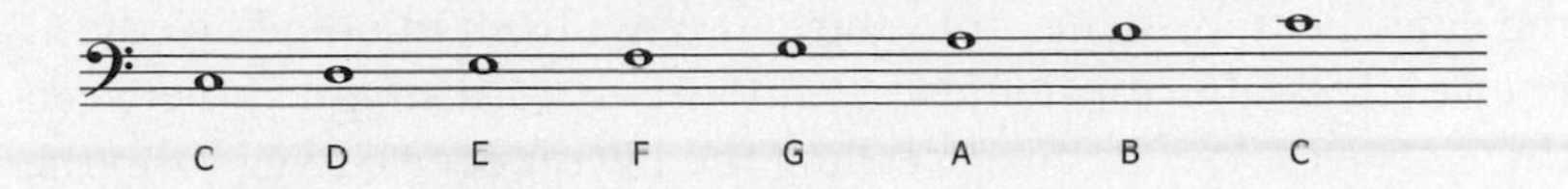

We represent the piano's black keys (sharps and flats) on our score as follows (for the treble):

This is basically the same for the bass.

I mentioned that there are several types of notes besides the whole notes I have written above. The different types of notes designate the timing of music. Music is written in various units of time: 4/4 or 4 beats to the bar (as in the fox trot), 3/4 or 3 beats to the bar (waltz), and so on. The whole note stands for 4 beats, the half note for 2 beats, and so on. The time for other types of notes are shown below.

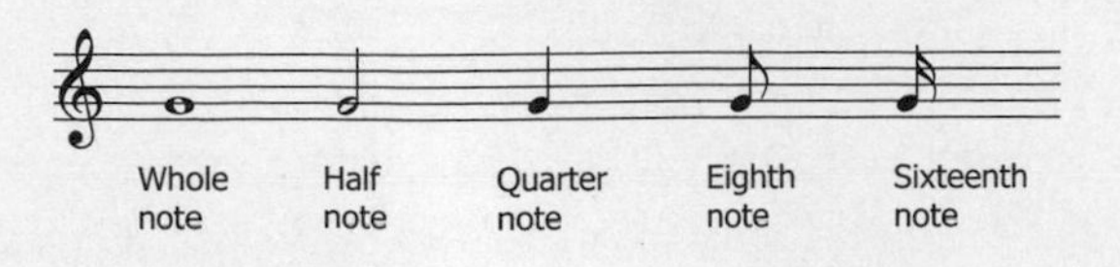

One of the other things you see are *rests.* Rests indicate places where no notes are played. Rests are designated in the following way.

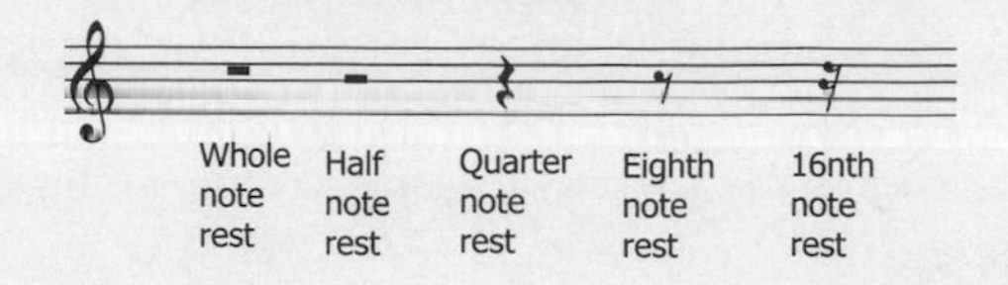

Also you will see a tie between notes. A tie between two notes means that the note is to be held (not replayed) and is designated by

Finally, throughout the book I will be referring to various octaves within the keyboard, where an octave is the interval from one note to the equivalent, or same, note above it (e.g., from middle C to the C twelve notes above it), and it is important to distinguish between these octaves. We do this by calling the lowest octave (e.g., from the lowest C on the piano to the B above it) C_1 to B_1. The next octave up is C_2 to B_2, and so on. In this scheme middle C is C_4.

There is, of course, considerably more to music than this, but these comments should be of help in understanding the music sections in the following chapters.

SOUND & SOUND WAVES

I

1

Making Music

How Sound Is Made

Music is sound, but it's a very special kind of sound. I think everyone would agree with that. In this chapter we'll be talking about how sound is produced and what properties it has to have to be music. Let's begin by defining sound. Sound is a wave that is created by a vibrating object; this object can take many forms, such as a tuning fork, the human voice, a siren, or a musical instrument. Once created, the sound propagates through a medium, usually air, from its source to another location where it is picked up by a receiver. The most common receiver is, of course, our ears.

Given that we're mainly interested in the form of sound that we call music, we have to distinguish it from the other sounds we hear. What is it that makes music different? There are many ways we can define it; a simple one is, Music is sound that is organized. We can also say that music differs from ordinary noise in that the vibrations associated with it are more uniform; in short, there are no sudden changes. Finally, musical sounds are, for the most part, pleasant and pleasing to the ear.

According to the first of our definitions, music is sound that is organized, so let's look at how it is organized. We see that it consists of notes, rhythms, phrases, and measures and has an overall form. All of these things help to organize it. Another important aspect of music is melody, in other words, the tune we whistle or hum after we've heard the music a few times. This melody is usually repeated several times throughout a musical piece and is something else that helps keep it organized.

The simplest form of music is the pure tone; it is the type of tone you get from a tuning fork. Pure tones are basic to music, but as we will see, they are not heard very often. Music composed only of pure tones would not be interesting.

So music is organized noise that has a melody and consists of structured rhythms and various types of pure tones. This is, of course, a very mechanical definition and doesn't convey what music really is and what it really does. As everyone knows, its most important role is to convey emotion, and it does this well. There's no doubt that it affects us all; it can convey joy, it can give us goose bumps, and it can even make us cry. Why does it have such a powerful force on people? This is one of the questions we will look at in this book.

The Motion of Waves

One of the first things we learn about sound is that it is a wave. This means that our study of sound will center on the study of waves. What exactly is a wave? Waves are, of course, all around us; we encounter waves of many different types every day. Besides sound waves, we have radio and TV waves, water waves, waves in our microwave ovens, and earthquake waves. In each case they are caused by some sort of vibratory motion.

One of the most familiar waves is a water wave, so let's begin by looking briefly at one. Assume you are sitting on the bank of a pond and throw a stone into the water. What do you see? When the stone hits the water, you will see a series of concentric rings that appear to move outward from the point where the stone hit, as illustrated in figure 5. Looking at these rings closely, you see that they consist of crests and depressions, or troughs. The tops of the crests are higher than the level of the water when there were no crests, and the bottoms of the trough are lower (fig. 6). The waves move out with a certain speed from where the stone struck the water, and it appears as if the water is actually moving. There is, indeed, some motion in the neighborhood of a particular crest or trough, but this consists only of a small amount of circular motion. The water as a whole does not move. The wave passes *through* the water.

Fig. 5. Girl throwing rock into pond, creating waves.

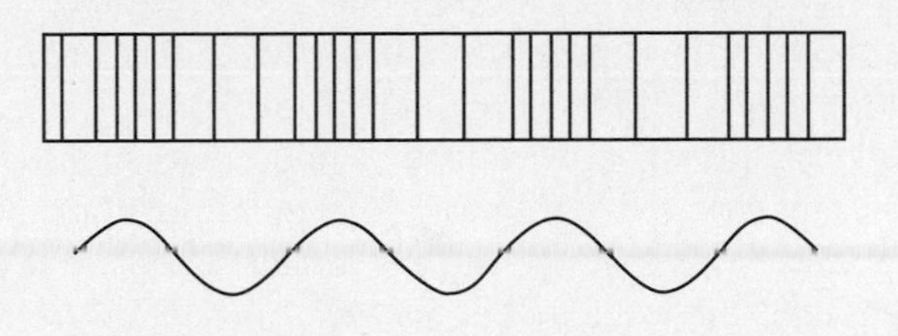

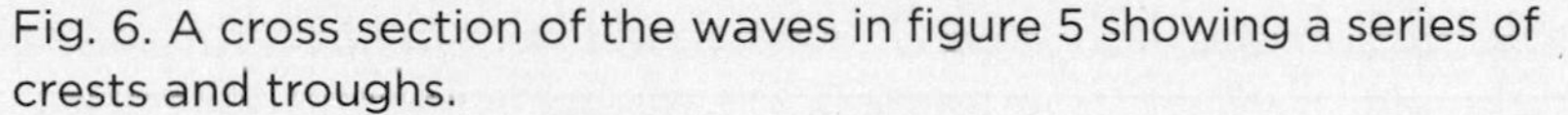

Fig. 6. A cross section of the waves in figure 5 showing a series of crests and troughs.

If you could take a cross section of the wave—in other words, cut it in a direction outward from the source—you would get a wiggly line that consisted of a series of crests and troughs. The curve of this line is known as a sine curve, since it is identical to the curve we get when we plot the trigonometrical function sine.

To understand this type of wave a little better, let's generate one and look closely at its properties. The best way to do this is to attach a rope to a doorknob or other projection and pull it tight, then give it a sudden upward jerk. A pulse that is similar to single wave (a crest and a trough) will travel down it from our hands to the knob (fig. 7). What we really want, though, is a series of these pulses. For this we have to keep jerking our hands up and down, and if we want the pulses to be equally spaced, we have to do it uniformly, or regularly. This will create an array of equally spaced pulses moving down the rope that will look exactly like the cross section we took of the water wave.

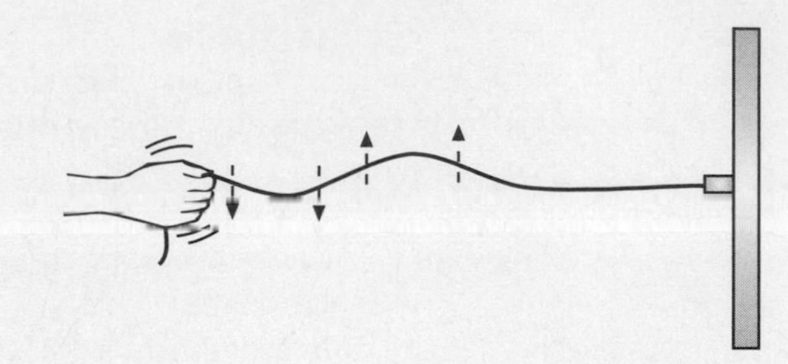

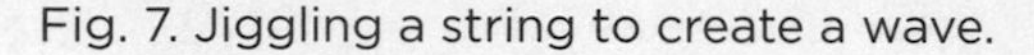

Fig. 7. Jiggling a string to create a wave.

The obvious conclusion from this is that vibratory motion is needed to create waves. As it turns out, though, a particular type of motion is critical in the case of music: *simple harmonic motion* (SHM). Simple harmonic motion is motion where the force on the object undergoing the motion is proportional to the displacement from its equilibrium position. It is said to obey Hooke's law.

A good example of something that undergoes simple harmonic motion is a taut string, like the string of a guitar, that is pulled to the side and released. It's easy to see that the farther we pull the string back, the greater the force pulling it back to its equilibrium position will be. So it obviously obeys Hooke's law. If we pull the string to the right and release it, as in figure 8, the restoring force will accelerate it back toward its equilibrium position, so that it moves faster and faster in this direction. You're no doubt familiar with acceleration in relation to your car. You have to accelerate to get up to speed; in other words, you have to increase, or change, your speed, so acceleration is *change in speed.*

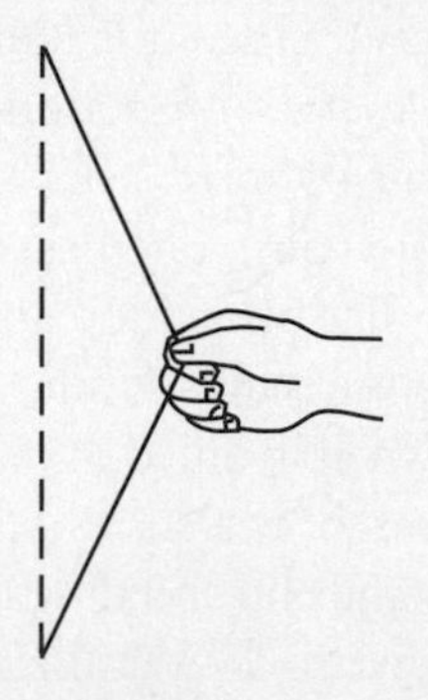

Fig. 8. When a string is pulled to the side and then released, it undergoes simple harmonic motion.

As the string approaches its equilibrium position (straight up and down in the diagram) its displacement from equilibrium decreases, and as a result, the restoring force also decreases. But it is the restoring force that is causing the acceleration; this is, in fact, known as Newton's second law of motion (acceleration is proportional to force). So the string moves faster and faster as it approaches the equilibrium position, but at the same time the acceleration itself is decreasing because the force is decreasing. Finally, at the equilibrium position, the restoring force is zero, and since the acceleration is proportional to the force, it is also zero. With no force acting on the string, it might seem that the string would stop moving, but it doesn't; in fact, it has its maximum velocity as it passes through the equilibrium position. Why doesn't it stop? Because of *inertia.* Inertia is something you experience every time you're in a car and it accelerates. Simply put, it is resistance to change in motion; an object will stay at rest or in motion until it is forced to do otherwise. A force will cause an object to move and accelerate, but to do this the force must overcome the object's inertia. Similarly, an object in uniform motion will not change its motion unless it is forced to do so. And in the case of our string, as it passes the equilibrium position there is no force on it. As it moves past equilibrium, however, the restoring force comes back into play, but now it's acting in the opposite direction. Since this force is proportional to the displacement of the string from equilibrium, it grows as the string continues to move. And, of course, associated with this force is an acceleration, but it is now in the opposite direction, so it's now a deceleration.

As the string continues to move to the left, the restoring force continues to increase, and because of the resulting deceleration, its velocity continues to decrease until finally it is zero. The string is now the same distance to the left of the equilibrium position as it was to the right when it started. At this point the restoring force again changes direction and is directed back to the equilibrium position. The string therefore begins accelerating in this direction, and as before, it passes through the equilibrium position with its maximum velocity and moves back to the position on the right from where it began. This process would continue indefinitely if there were no air resistance or friction at the ends of the string. But in practice there is always some friction, and as a result, the vibrations are damped and eventually die out.

It's obvious that simple harmonic motion is relatively complicated. The reason is that the velocity of the object undergoing the motion is continually changing, and so is its acceleration. The changes are always smooth, though, so there are no abrupt motions.

Stretched strings are not the only things that undergo simple harmonic motion when they are pulled aside. Pendulums also have this property (fig. 9). Galileo was sitting in a cathedral one day in 1583 when he noticed the chandeliers swinging back and forth. They were all of the same length, but they were not all swaying the same distance from their equilibrium positions (straight up and down). We refer to the distance from the equilibrium position as the *amplitude*. Galileo noticed that, regardless of their amplitude, the chandeliers appeared to move back and forth in the same length of time (we refer to this as their *period*). He used his pulse to time them, and sure enough, their period was the same.

The observation fascinated Galileo, and when he returned to his home he decided to look into the motion of pendulums further. He attached a bob to the end of a string and began experimenting with it. The first thing he noticed was that the period of the swing did not depend on the weight of the bob, but it did depend on the length of the pendulum. In fact, it varied as the square root of the length. This meant that if a pendulum was one foot long and had a period of one second, a pendulum two feet long would have a period of $\sqrt{2}$ seconds, and a pendulum four feet long would have a period of $\sqrt{4} = 2$ seconds.

Now, let's look at the motion as we did in the case of the stretched string. In this case when the bob is pulled to the side, a restoring

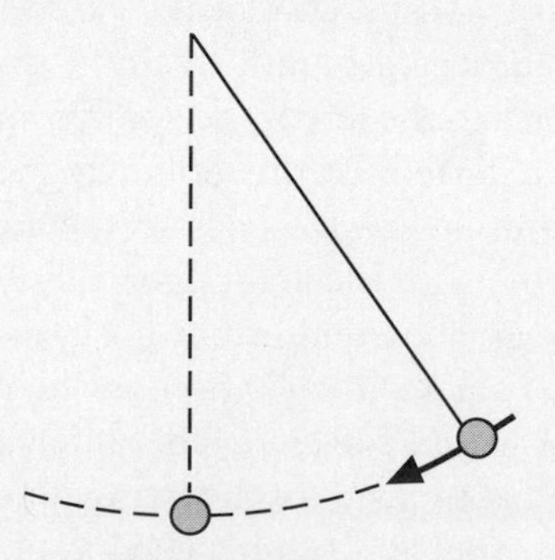

Fig. 9. A simple pendulum.

force also acts on it, but this case is different because the bob moves in an arc of a circle when it returns to equilibrium. At its extreme it is raised to a higher level as compared with its equilibrium position. Because of this, when it is released, gravity acts on it, pulling it downward in an arcing curve. But it is obvious that gravity is not the only force involved. There is a force on the string—an upward pull. The gravitational pull that causes the restoring force is therefore only the part that is not balanced by the upward pull of the string.

Because the gravitational force is largest when the bob is first released, the acceleration is also the greatest at this point. As the bob drops, however, the gravitational force decreases and so does the acceleration. But because of the acceleration, the velocity of the bob increases and is at a maximum when the string is vertical. At this position, there is no unbalanced force, but because of inertia, the bob swings through this position and past it. On the other side of this position the force acts back toward it, and the bob therefore slows down. Eventually it stops at roughly the same distance on the opposite side of the vertical. At this position the gravitational force has built up again, and the bob starts back down toward the equilibrium position (the vertical). It performs this motion over and over again, and if it were not for friction it would continue to do so indefinitely.

It's easy to see that this motion is very similar to the plucking of a string. Both, in fact, are simple harmonic motion.

Types of Waves

Simple harmonic motion is basic to wave motion. One way to see this is to go back to our rope tied to a doorknob. We saw that we could create a series of equally spaced pulses that moved down the rope and that they were similar to the cross section of the waves we saw on water. Looking at this wave closely, we see that it consists of equally spaced *crests* and *troughs*. The crests are the places where the rope is displaced above its usual equilibrium position (when it is pulled tight), and the troughs are the valleys created below the equilibrium position. There are, of course, places along the rope where it is not displaced from its equilibrium position; they are referred to as *nodes*. This type of wave is called a *transverse wave*. It is the type of wave that is set up on a violin string, or a string in a piano. The motion of the rope is perpendicular to the motion of the wave in this type of wave; it is, in fact, a basic property of a transverse wave.

There is another type of wave that is also important in nature. To see how this type of wave is generated it is helpful to use a Slinky. I'm sure you are familiar with this toy; you probably played with one when you were young. It is a continuous metal coil that can easily be stretched out. If you attach a Slinky to a doorknob, pull it out straight (or at least as close to straight as you can), and then hit or pound the end of the Slinky, you will, as in the case of our rope example, see a pulse move along it to the doorknob. But this pulse is different from the one with the rope (fig. 10). It's a disturbance caused by the back-and-forth movement of the coils in the Slinky. The first coil is disturbed by the blow you gave it; it pushes the second coil and displaces it from its equilibrium position. This causes a push or pull on the third coil, which is displaced from its equilibrium position, and so on. The result is a disturbance that moves down the Slinky. The disturbance in this case is in the direction the wave is traveling, so in this respect it is different from a transverse wave. In several respects, however, this wave is similar to the transverse wave we talked about above. It is referred to as a *longitudinal wave*, and as we will see, it is also of importance in music.

Let's look at both of these waves in more detail, beginning with the transverse wave. We saw that it has crests and troughs and has a regular repetitive shape down its length. And since it has the same shape as the trigonometrical sine function we call it a sine wave. In the sine wave a small section of it is repeated over and over. The distance of repetition (from one point to a similar point farther on) is known as the *wavelength* of the wave, which is usually designated by a lowercase Greek lambda, λ. The wavelength can be measured from

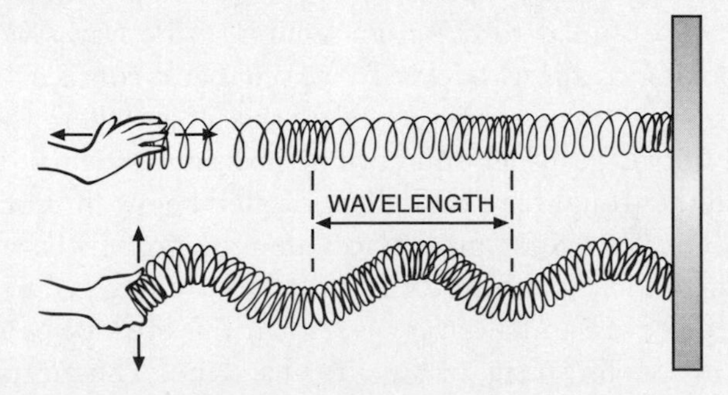

Fig. 10. Using a Slinky to illustrate waves. Both transverse and longitudinal waves are shown.

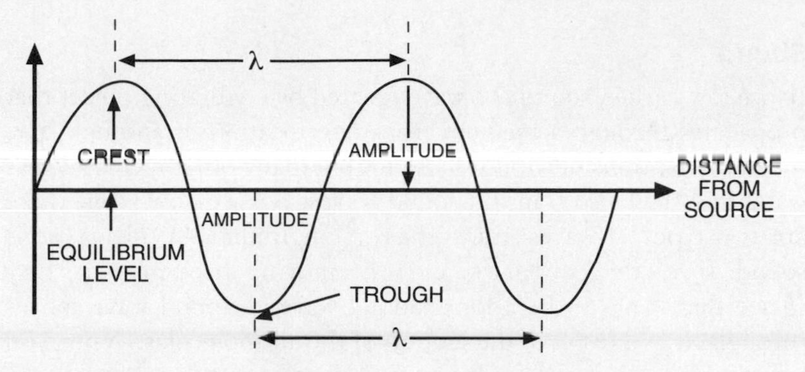

Fig. 11. An illustration of amplitude and wavelength (λ).

the maximum of one crest to the next, or from a minimum of a trough to the next, or any other two equivalent points. Similarly, the distance from the equilibrium position to the maximum of a crest is called the *amplitude* of the wave. Wavelengths and amplitude are shown in figure 11. Another important property of the transverse, or sine, wave is the number of crests (or troughs) that pass a given point per second; this number is referred to as the *frequency* of the wave and is usually designated by *f*.

If we have two waves traveling down two ropes that are side by side and the crests and troughs from one line up with the other, we say that the two waves are *in phase*. If the crests and troughs do not line up the two waves are *out of phase*. Furthermore, we can specify how far they are out of phase; if, for example, a crest lines up with a tough on the other wave, we say they are out of phase by half a wavelength.

Let's turn now to the longitudinal waves we saw on the Slinky. As we look down the wave we see regions where the coils of the Slinky are closer together than usual; they are referred to as *compressions*. We also see regions where the coils are farther apart than usual; these regions are referred to as *rarefactions*. Compressions and rarefactions are analogous to crests and troughs in transverse waves. In the same way, therefore, we have a wavelength for a longitudinal wave; it is the distance between two rarefactions or two compressions (or other equivalent points). In addition, the points that remain at equilibrium correspond to nodes, and the number of compressions (or rarefactions) that pass a given point per second is the frequency of the wave.

Sound

As we saw earlier, sound is a wave created by a vibrating object that propagates through a medium from one location to another. The medium that transmits it is usually air, but many other media such as water and steel also transmit sound waves. But we now know there are two types of waves: transverse or longitudinal. Which type is sound? If you think about a sound wave moving through air, it's easy to see that it has to be a longitudinal wave. A sound wave moves through air as a result of the motion of the air molecules. When you talk or sing, your vocal cords exert a force on the air molecules next to them. As a result, these molecules are displaced from their equilibrium position. They, in turn, exert a push or a pull on their neighbors, causing them to be displaced from their equilibrium position. These push and pull motions, continuing all the way to the receiver, are like the waves that traveled down the Slinky.

Sound waves need a medium such as air to propagate them, and as a result they are referred to as mechanical waves. There are also nonmechanical waves that do not need a medium to transport them. They are called electromagnetic waves, and as we will see, they also play an important role in music. A radio wave is an example.

Properties of Sound Waves

A tuning fork is a familiar device that creates a sound wave of a single frequency. As the tines, or prongs, of the tuning fork vibrate back and forth they push on the air molecules around them. The forward motion of the tine pushes molecules together creating a compression; then, as the tine moves back it creates a rarefaction. If we place an open tube next to the tuning fork, compressions and rarefactions will be set up in it, as is shown in figure 12. This wave has the same properties as the longitudinal wave we talked about earlier. The distance between successive compressions (or rarefactions) in the tube is the wavelength of the sound wave. Tuning forks are made to vibrate at a particular frequency. Thus, if a tuning fork is designed to sound middle C, it will vibrate at 256 vibrations/sec. The frequency of the longitudinal waves passing down the tube would therefore also be 256 vibrations/sec.

Another way to look at this tube is to measure the air pressure at each point along it. At a compression, this pressure would be higher

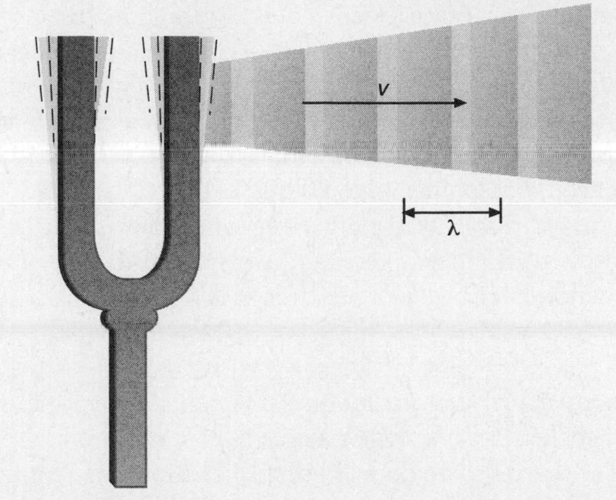

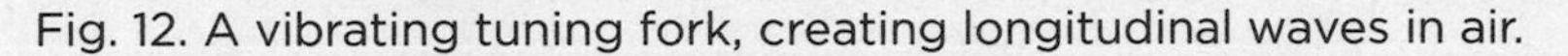

Fig. 12. A vibrating tuning fork, creating longitudinal waves in air.

than normal (equilibrium pressure), and at each rarefaction it would be less than normal. We could, in fact, make a plot of pressure versus time for the tube. Such a plot is shown in figure 6; we see immediately that it looks like a transverse wave—in other words, it's a sine wave. This shows the close relationship between transverse and longitudinal waves. This is not to say, of course, that they are the same. They are distinctly different.

We've seen that a sound wave has a particular frequency, which is the number of compressions that pass a given point per second. And for years we referred to the units of frequency as so many vibrations/sec, but we now use a unit called the Hertz (Hz), where 1 Hz = 1 vibration/sec. The sound emitted from the tuning fork mentioned above is therefore 256 Hz.

Another important characteristic of sound is its *period.* Its period is the time for a compression (or rarefaction) to move between two successive equivalent points. The relationship between period and frequency is frequency = 1/period—that is, the frequency is the reciprocal of the period.

Audible sound covers a relatively large range of frequencies. The human ear is capable of detecting frequencies from about 20 Hz up to 20,000Hz. Sounds below this range are usually referred to as *infrasound,* and sounds above it are called *ultrasound.* The upper range of audible sound actually varies considerably for humans. As you get older, for example, your upper range decreases, and you can't hear sounds anywhere near 20,000 Hz. Animals generally can hear a wider range of frequencies than humans. Dogs, for example, can hear frequencies of 50 Hz up to 45,000 Hz. Bats can detect frequencies up to 120,000 Hz, and dolphins, up to 200,000 Hz.

When most people hear a high frequency, they say it has a high *pitch*. And pitch is, indeed, generally considered to be synonymous with frequency, but as we will see later, it is not exactly the same.

Intensity of Sound

Another property of sound is loudness. When a wave passes through a medium such as air, it transports energy as it moves, where energy is defined as the ability to do work (in units of joules). This energy is transmitted to the medium by the vibrations that create the wave. It depends on the amplitude of the vibration: the greater the amplitude, the greater the energy, and the louder the sound. In the case of a guitar string, for example, the farther it is pulled to the side, the greater is its amplitude, and the louder is the sound it produces.

Loudness is associated with a quantity called *intensity*, where intensity is defined as the amount of energy that is transported past a given area of a medium per unit time. But energy per unit time is power, so intensity is power per unit area, and since the units of power are watts, the units of intensity are watts/meter^2 (W/m^2).

As a sound wave spreads out in a medium, its intensity decreases. This is easy to see by looking at the diagram in figure 13. As the wave moves outward, the same amount of energy is spread over the area at 1 m, 2 m, 3 m, and so on. Since the area is bigger in each case, and the

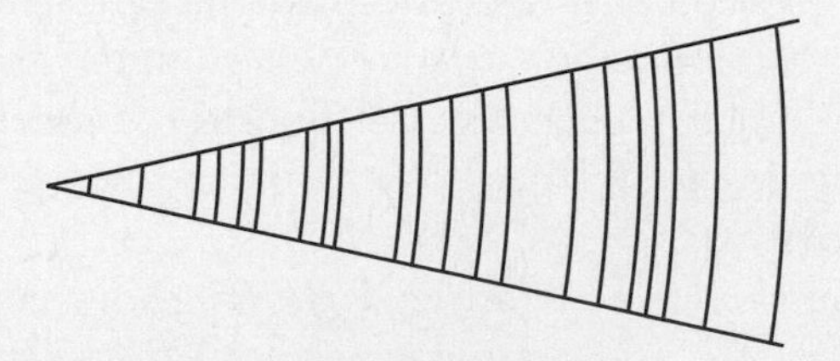

Fig. 13. The "spreading out" of a wave.

energy is the same, the energy per unit area is less. The relationship in this case is called an inverse square law, which means that the intensity of the wave drops off as the square of the distance from the source. Therefore, if the distance is doubled, the intensity will decrease by a factor of 4; if the distance is tripled, it will decrease by a factor of 9; and so on.

The range of sounds that impinge on the human ear each day varies considerably. It is so large, in fact, that physicists use a scale for intensities that is based on multiples of 10. This logarithmic scale is referred to as the *decibel scale*. A sound that is 10 times as intense as our threshold of hearing (the threshold being something we can barely hear) is said to have a sound level of 10 decibels (dB). In mks units (measured as meters, kilograms, and seconds) this is 10^{-12} W/m^2. A sound that is $10 \times 10 = 100$ times as intense has an intensity of 20 dB (or $10^{-11} W/m^2$), and so on. So a sound of 10 dB is 10^1 times threshold, a sound of 20 dB is 10^2 times threshold, a sound of 30 dB is 10^3 times threshold, and so on. Table 1 gives examples of the intensity of some common sounds.

The intensity of sound is something that can be measured exactly, and it is obviously associated with loudness, but it is not exactly the same. The loudness of a sound actually depends on several factors. All people do not hear a given intensity as equally loud. Older people, for example, do not hear a particular intensity to be as loud as a younger person. Also, the frequency of the sound has an effect. Different frequencies of the same intensity are perceived to have a different loudness.

Table 1. Intensity levels of some common sounds

Sound	Intensity level (dB)
Rustle of leaves	10
Average living room (background noise)	40
Normal conversation	60
Street traffic in large city	80
Large orchestra (playing moderately loud)	95
Factory floor	100
Threshold of auditory pain	120
Takeoff of jet airplane	140
Perforation of eardrum	160

The Speed of Sound

When a sound wave passes through a medium such as air, particles are disturbed, which in turn, disturb adjacent particles so that energy is transported through the medium. The individual molecules do not move very far, and the overall medium does not move at all, but the wave passes through it with a certain speed that depends on several factors. It's well-known that speed is defined as distance divided by time, so it has units of m/sec in mks units (or ft/sec in British engineering units).

Two properties determine the speed of a wave through a medium: the medium's *inertial properties* and its *elastic properties.* When we talk about inertia in relation to sound, we are referring to the inertia of the particles that make up the medium. Particles with greater mass have greater inertia and are therefore less responsive to a wave passing through them. This means that, in general, the greater the density of the material, the slower the wave will travel. Therefore, it will travel faster in light gases than it will in a heavy or dense gas (assuming other factors are the same).

Elastic properties are those related to how well a material retains its shape when subjected to a force. Steel is a good example of a material that is very rigid and inelastic; rubber is a material that is not. High rigidity and inelasticity are related to the strength of the atomic or molecular forces within the material, and since metals generally have stronger molecular forces within them than fluids, and fluids have stronger molecular forces than gases, the speed of sound is therefore highest in solids, next highest in liquids, and lowest in gases.

Temperature and pressure also have an effect on the speed of sound; it increases with increasing temperature or pressure because both elasticity and inertia are affected. At normal pressure and a tem-

Table 2. Speed of sound through various media at 0°C

Transmission medium	m/sec	ft/sec
Air	332	1,087
Hydrogen	1,270	4,167
Water	1,450	4,757
Iron	5,100	16,730
Glass	5,500	18,050

perature of 0°C, sound travels at 331.5 m/sec (1,087 ft/sec), and at 20°C (68°F) it travels at 343 m/sec (1,130 ft/sec). Table 2 gives some examples of the speed of sound in various substances.

The velocity of all waves, including sound waves, is related to their wavelength and frequency by

velocity = wavelength × frequency,

or in formula form,

$$v = \lambda f.$$

This formula does not imply that different wavelengths or frequencies of sound travel at different velocities. The speed of sound does not depend on either of these quantities. A change in wavelength does not change the velocity: it changes the frequency, and vice versa. The speed of sound depends only on the properties of the medium it is moving through.

2

The Sound of Music

Perception

Imagine that you are sitting in a concert hall waiting in anticipation for the concert to start. All at once the lights go down and colored lights start flashing across the stage. The musicians run out waving. A tingle of excitement surges through you as the first notes sound. You lean back in your seat as the musicians begin to play. The beat pulses through the crowd, who soon have their arms in the air clapping to the beat.

I'm sure most of you don't need to imagine it. You've been there. And you'll agree that music brings a lot of excitement and joy into our lives. Of course, what allows us to enjoy it is our sense of hearing. Surprisingly, though, few of us ever stop to think about how important our hearing is, and how or why it works. In this chapter we will see that it is, indeed, a miraculous process. We will look at how music is processed by the ear, in particular, how it distinguishes various pitches and degrees of loudness. At this stage, however, we will be considering only pure tones—in other words, tones of a single frequency. We saw earlier that pure tones are rarely heard in music; nevertheless, they are important in understanding more complex tones.

Overview of the Ear

The ear consists of three parts: the outer, middle, and inner ear (fig. 14). Each of them plays an important role in the processing of sound,

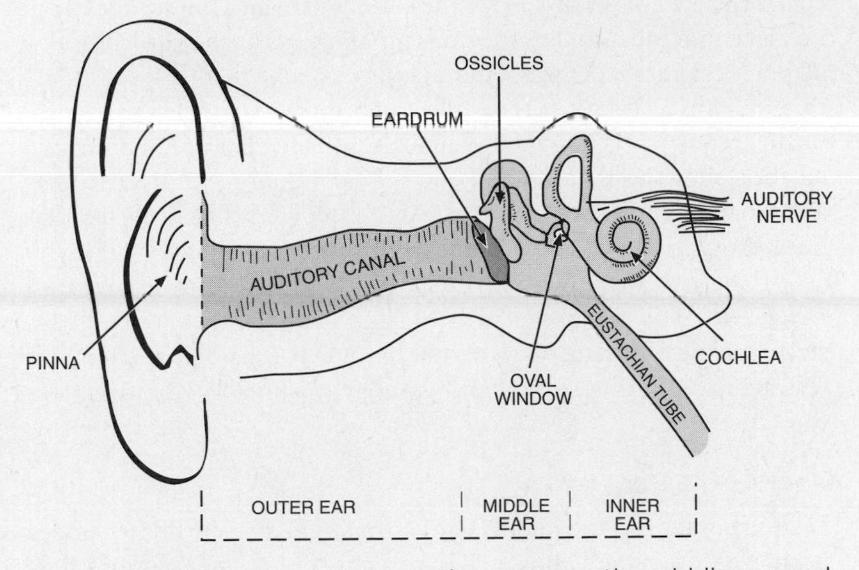

Fig. 14. The three parts of the ear: the outer ear, the middle ear, and the inner ear.

and we will look at each in detail, but first let's review what we know about the generation of a sound. We can assume it is coming from a musical instrument or perhaps from someone's voice. We know that a pressure wave is created in the air—a longitudinal wave in which the air molecules are spread out into regions of compression and rarefaction. It travels at 343 m/sec toward our eardrums, and if it is in the frequency range 20 to 15,000 Hz, our auditory systems will respond to it. In essence, it will cause our eardrums to vibrate, and they, in turn, will transfer the vibrations through the middle ear to the inner ear. As we will see, the inner ear is the most complex part of the ear. This is where the vibrating signal is changed to an electrical impulse that is transmitted to the brain. The inner ear is deep in the skull for a reason: it's the most sensitive part of the ear and the part that is most difficult to correct if anything goes wrong, so it has to be protected.

The Outer Ear

The outer ear is the visible part of the ear (or at least some of it is visible) and the part that most people refer to as the "ear." The external portion, or ear flap, is referred to as the pinna, or auricle. Its major role is to help collect and direct the sound wave into the auricle channel, a channel that is about 3 cm long and ends at the eardrum. For all practical purposes the auricle channel is a tube that is closed at one end, and it is well-known that tubes such as this have a natural or resonant frequency. (I'll say more about resonance later.) A simple calculation shows us that a closed end tube of about 3 cm in length has a resonant frequency of approximately 3,000 Hz. This means that our hearing should be most acute in this region, and indeed it is.

At the end of the auricle channel is the eardrum, or *tympanic membrane*. It is composed of a fibrous material that looks like a stretched piece of skin. Most people assume it is flat like a drum, but it is actually slightly conical and is most sensitive to vibration near the center.

The Middle Ear

Like the outer ear, the middle ear is surrounded by air, so it is under normal air pressure. Unless you have a cold or are undergoing a sudden change in elevation, the middle ear is always at the same air pressure as the outer ear. This is a result of the Eustachian tube, which connects the middle ear to the throat (fig. 15). You are no doubt familiar with the feeling of pressure on your eardrums when you are in an airplane and it takes off, or when you dive into deep water. Furthermore, you know that if you swallow, the feeling of pressure goes away. In effect, what is happening is that the Eustachian tube is opening and equalizing the pressure between the middle and the outer ear.

Any vibrations that impinge on the eardrum are transferred inward by a mechanical system of levers made up of three bones called *ossicles*. They are the hammer (*malleus*), the anvil (*incus*), and the stirrup (*stapes*), also shown in figure 15. The hammer is attached to the center of the eardrum and tightly bound to the anvil by ligaments, so that when the hammer moves, the anvil moves in unison with it. The anvil, in turn, is attached to the stirrup, and the footplate of the stirrup is attached to a small window called the *oval window* that leads to the inner ear.

In many respects both the eardrum and the oval window are like the diaphragm of a microphone. As you probably know, because of the variations in air pressure caused by your voice, the diaphragm vibrates when you speak into a microphone. The diaphragm, in turn, is attached to a device that creates an electrical current that is pro-

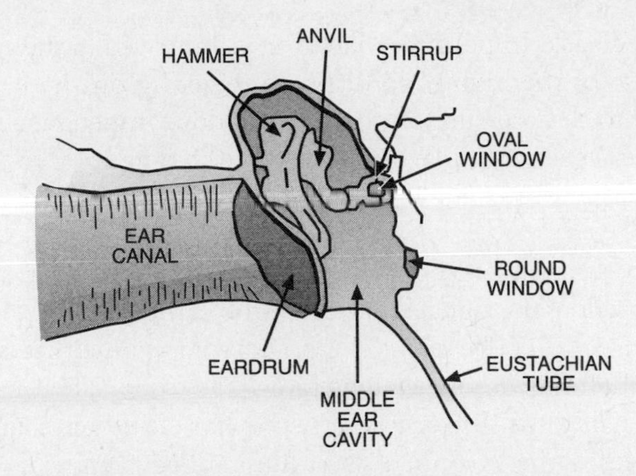

Fig. 15. The middle ear, showing the hammer (*malleus*), the anvil (*incus*), and the stirrup (*stapes*).

portional to its movement. In effect, it picks up air pressure vibrations and changes them into an electrical signal.

In the middle ear, the hammer, the anvil, and the stirrup act as a simple lever, where the motion of the hammer and the anvil causes the stirrup to push in and out on the oval window. This creates a wave in the fluid on the other side of the window. The signal is increased slightly by the lever action, but the motion is not actually amplified; rather, the overall system increases the force of movement by a factor of about 1.3. We say it has a mechanical advantage of 1.3. This might not seem like much, but it is nevertheless important. The real amplification, however, occurs because of the difference in size of the eardrum and the oval window. The area of the eardrum is about 0.6 cm^2, while the area of the oval window is only 0.035 cm^2. This is a ratio of about 1:20, so overall, the pressure exerted on the eardrum is increased by about 25 as it passes through the oval window. This is a significant amplification; it actually allows us to hear sounds 600 times weaker than we could without this amplification.

Another important role of the middle ear is that it produces much better *impedance matching* between the eardrum and the oval window. This is important at any boundary between two different media. Impedance is a measure of how much alternating force must be applied to produce a particular vibrational velocity in a second medium. If the impedance of the first medium matches that of the second, all or most of the energy will be transmitted; if not, some of the energy will be reflected back. Impedance matching is improved in the case of the ear because of the high ratio of pressure acting on the eardrum as compared to the pressure on the oval window. Impedance matching in this case is about 50–70% of perfect in the range 300 to 3,000 Hz.

The Inner Ear

The inner ear is the most complicated of the ear's three regions. It is the region where the signal is changed from a mechanical vibration to an electrical signal. The major component of the inner ear is the *cochlea*. From the outside the cochlea looks like a snail shell with about two and a half turns. It consists of two major chambers that are connected; within the chambers is a fluid called *perilymph* (fig. 16). (There is a chamber between them that is filled with a different fluid called *endolymph*, but it is relatively small and we will ignore it in our discussion.)

To understand how the sound wave is transferred into the cochlea, it's best to imagine the cochlea stretched out. When stretched, it is about 3 cm long, and the two chambers are separated by a narrow strip of skin called the *basilar membrane*. The top section (inside the oval window) is called the *scala vestibuli*, and the bottom is called the *scala tympani*. The narrow basilar membrane is stretched tightest near

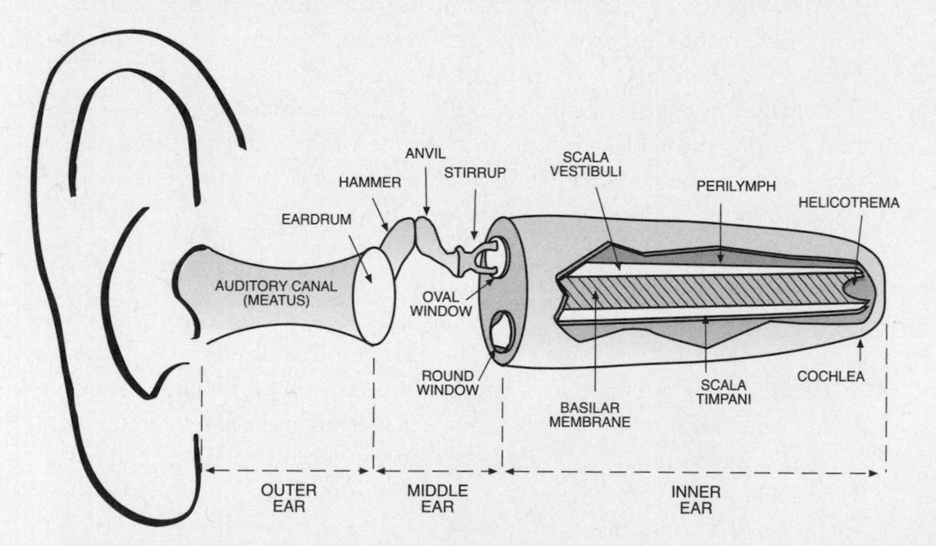

Fig. 16. The inner ear, showing the basilar membrane, the perilymph, and the scala tympani.

the oval window; it widens with distance from the window, so it is generally triangular in shape. At the far end of the inner ear is the *helicotrema*, which connects the two chambers. Near it, the strings of the basilar membrane are under much less tension and therefore vibrate at much lower frequencies.

The underside of the basilar membrane is coated with about 30,000 hair cells arranged in rows. Protruding from each of them are about 12 to 40 hair cilia that look like a small tuft of hair. Finally, over the cilia is a soft pad called the *tectorial membrane*. Auditory nerves are connected to the hair cells; they consist of a bundle of fibers that go to the brain.

The vibrations at the oval window are transferred to the perilymph fluid, causing a disturbance in the fluid that moves along the basilar membrane, affecting the hair cells; they in turn produce electrical impulses that travel to the brain and are interpreted by the brain as noise.

Recognizing Pitch

One of the most important functions of the inner ear is its ability to recognize different pitches, or frequencies. How does it do this? Before I answer this, let's look at the concept called *resonance*, which is involved in recognizing pitches. As I mentioned earlier, all objects have natural frequencies—in other words, frequencies where they oscillate most efficiently. If a periodic force is applied to an object, the object will vibrate at the frequency of the applied force. If this frequency is not the object's natural frequency, however, the forced vibrations will occur at the expense of considerable energy—energy that is used to overcome the natural frequency. If the forced frequency matches the natural frequency, on the other hand, the object will vibrate much more easily, and the oscillations will build up. In particular, if you place two tuning forks of the same frequency close to one another and set one in motion, the second one will soon pick up the vibration. Energy has, in effect, been transferred from the first tuning fork to the second.

The first theory of how the ear recognizes pitch was put forward by the German physicist Hermann von Helmholtz. Although his theory is now considered to be an oversimplification, it is worth considering because it is correct in several respects. Helmholtz was both a physiologist and a physicist. He studied medicine at his father's insistence but actually preferred physics. After working as a surgeon in the Prussian army for several years, he obtained a position as professor of physiology at the University of Königsberg. (He later taught at Heidelberg and Berlin.) He was tremendously interested in the mechanism of the ear, but he also had an intense interest in the function of the eye and made important contributions to both fields. His interest in the ear no doubt came about as a result of his interest in music; he was an accomplished musician and later applied many of the principles of physics to music.

Helmholtz's major contribution to hearing was in connection with the basilar membrane. If you look down it from the oval window it has a triangular shape and resembles a dulcimer (or harp) with short, tight strings near the oval window and long, slack strings at the other end. When a pure tone strikes the oval window, it sets the window vibrating and this vibration is transferred to the perilymph fluid. Helmholtz postulated that somewhere along the basilar membrane there was a string with this frequency. In other words, it was resonant at this frequency and would therefore pick up this frequency from the fluid and begin vibrating. Other strings would not. Directly beneath this string was a hair cell, and the resulting vibration would cause it to fire an electrical impulse that would travel to the brain.

Helmholtz's ideas seemed a little too simple to Georg von Békésy, a communications engineer in Budapest, Hungary, who worked for the Hungarian telephone system. In 1928 he was studying the relationship between the mechanism of hearing on the telephone and the mechanism of human hearing when he began to wonder how much better the human ear was than the mechanism of the telephone. As a result, he decided to study the ear. He began by building mechanical models of the cochlea using a tube filled with water. Down the middle of the tube he placed a tightly stretched membrane, which represented the basilar membrane; at the nearby end he placed another membrane representing the oval window. When the oval window was set vibrating, he noticed that a wave swept along the membrane; by adjusting the tension along it, he found that he could confine the major bulge of the wave to a particular region of the membrane. In other words, the amplitude of the wave increased significantly at only one position along the basilar membrane.

Békésy went on to experiment with the cochlea of animals, and later with those from corpses. Using several ingenious techniques, he was able to see that a wave, similar to the one he had seen in his model, swept down the basilar membrane when the oval window vibrated. And as in his model, the waves had a maximum amplitude at a particular position along the membrane. In short, the wave's amplitude increased as the wave moved along the membrane until the amplitude reached a maximum at a particular point and then fell off rapidly (almost instantaneously). The point at which the wave reached its maximum corresponded to the frequency of the sound detected by the ear. At this point the hair cells responded, sending a message to the brain. The electrical impulses from a given hair are all of the same strength, so that for a louder sound, more pulses have to be sent.

Loudness and Loudness Curves

We had a brief introduction to loudness in the last chapter, but there are several aspects of it that we did not cover. We will look at them in this chapter. As we saw, the usual measure of loudness is intensity, a physical quantity that can be measured directly (in W/m^2). We also introduced a unit called the *sound level* (SL) that was measured in decibels (dB). Because the range of sound intensities is so large—approximately a trillion (10^{12})—decibels are measured with a logarithmic scale ranging from 0 to approximately 120.

In reality, the loudness of a sound is personal. What is loud to one person may not be particularly loud to another, so loudness is subject-dependent. In addition, the ear does not hear all frequencies the same. If you use a signal generator to generate a sound level of, say, 40 dB at 500 Hz and have somebody listen to it, then produce a sound of 40 dB at 1,000 Hz, they are likely to say that the two sounds have different loudness. It is convenient, therefore, to have a subjective system of measure. For this we need a group of people listening to a particular sound level at several different frequencies, adjusted so that they all seem to have the same loudness. A reference frequency is taken at 1,000 Hz. This means that the group listens to, say, 40 dB at 1,000 HZ, then listens to a tone at 500 Hz which is gradually adjusted so it seems to have the same loudness. Then they do the same thing at 300 Hz, 200 Hz, and so on. We can plot the results in a graph as shown in figure 17. The reference level is, of course, a straight line at 1,000 Hz.

From the graph we see that at 100 Hz a tone level of 30 dB sounds to the group as if it has a loudness of 50 dB. Similarly, the same sound

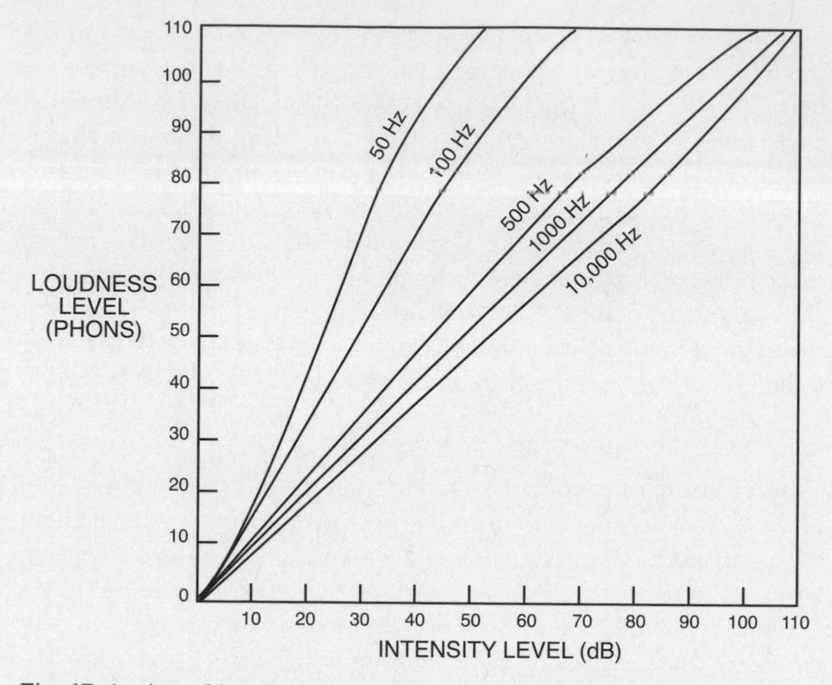

Fig. 17. A plot of loudness level versus intensity in decibels (dB).

level at 10,000 Hz sounds as if it has a loudness of 15 dB. In general, low frequencies sound louder than they actually are, and high frequencies sound less loud. On the basis of this we can define a new loudness measure with units of *phons* called the loudness limit (LL). By definition, a phon is a subjective measure of loudness intensity where 1 phon is equal to 1 dB at 1,000 Hz, but differs at all other frequencies. The major difference between the two scales occurs at low frequencies; at high frequencies the difference is relatively small.

We can also plot this difference to give what are called *equal loudness curves.* In figure 18 we plot sound level in decibels versus frequency in Hertz where curves of equal loudness are plotted.

Pitch Acuity and Just Noticeable Difference (JND)

As in the previous section we again deal with pure tones, and we ask, What is the smallest pitch (or frequency) interval that can be distinguished? In other words, when one tone is played and then another of very close pitch, can you say with certainty that one is higher than

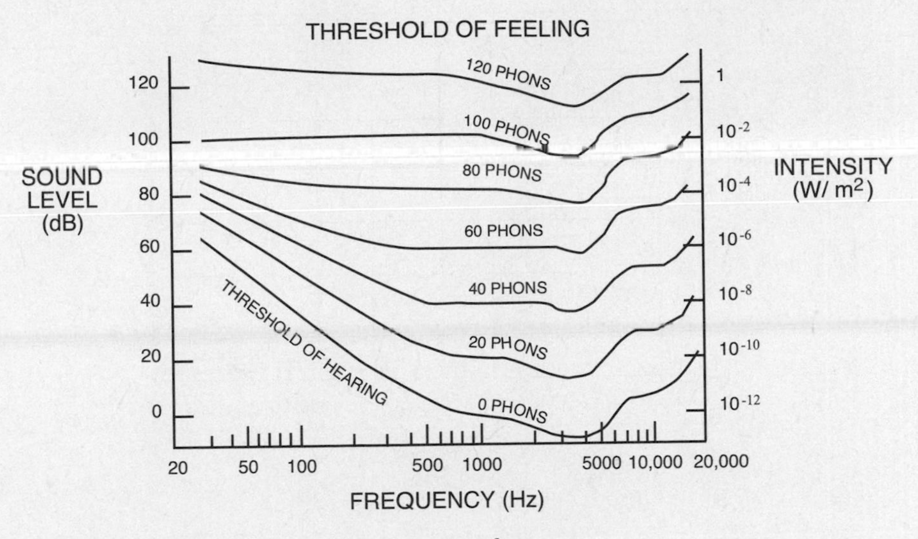

Fig. 18. A plot of sound level versus frequency.

the other? This is referred to as pitch acuity, or pitch difference sensitivity, and is usually measured as a percentage.

In the case of pitch, a pure tone is changed at a constant sound level until the subject says there is a difference. The difference between the initial frequency, f_1, and the final one, f_2, is $f_1 - f_2$. Since this difference is quite sensitive to frequency, we have to include the average of f_1 and f_2, which we will call f. We then define $(f_1 - f_2)/f$ in percentages as the pitch acuity. A typical value is 0.5 to 1%. If we give only the difference in frequency in the two tones, we refer to it as the just noticeable difference (JND).

The results can be summarized by saying that acuity is poorest at high frequencies; see figure 19. A typical value in the frequency range 500 to 5,000 Hz is 0.5 to 1%, but it is much larger below 500 Hz. In the same way, JND is approximately 1 Hz for all frequencies below 1,000 Hz and begins to increase above that. Beyond 5,000 Hz the rise is rapid, and our pitch judgment becomes poor. Beyond 10,000 Hz all ability to distinguish pitch vanishes.

We can make a similar plot with sound levels (e.g., using slightly different sound levels and the same frequency), as seen in figure 20. The subject hears one sound, then a slightly different sound level at the same frequency and is required to say which is the loudest. Again,

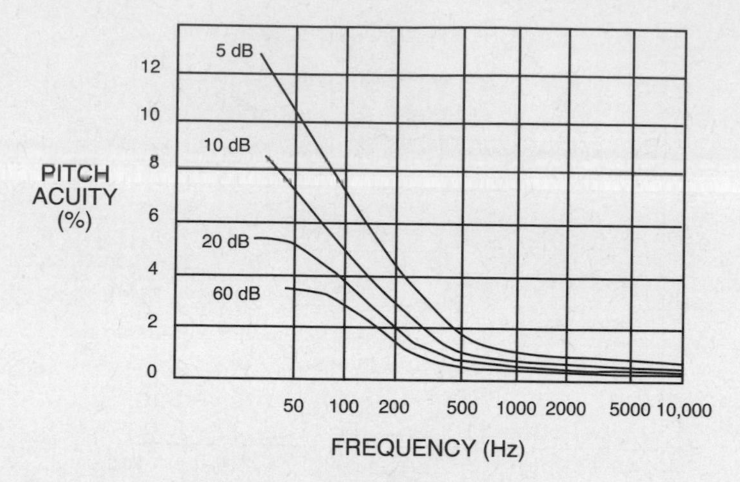

Fig. 19. A plot of pitch acuity versus frequency.

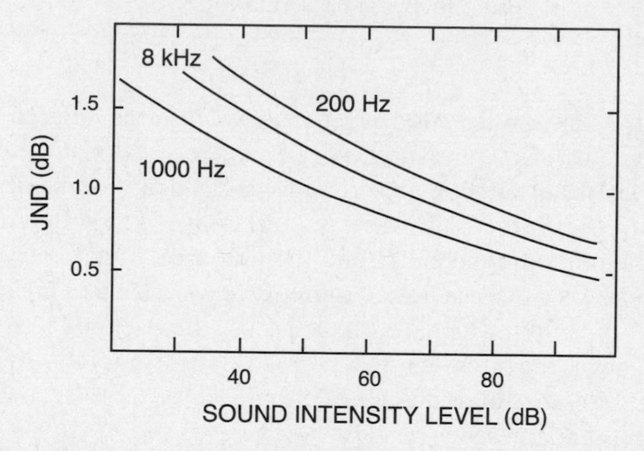

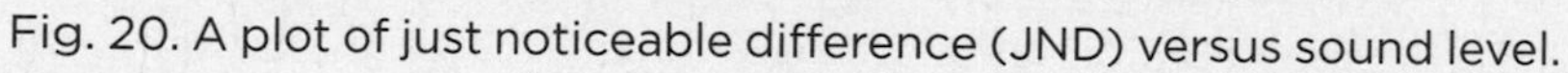
Fig. 20. A plot of just noticeable difference (JND) versus sound level.

the results depend on the frequency. The JNDs tend to be a little larger at low frequencies and lower intensities. In general the JND for sound levels is between 0.5 dB and 1 dB.

Hearing Loss

Hearing loss can be caused by several things, but one of the greatest hearing losses results from continued exposure to very loud intensities, particularly those over 100 dB.

Two basic kinds of deafness occur, referred to as *conduction deafness* and *nerve deafness.* Conduction deafness occurs when the sound is not properly conducted from the eardrum to the inner ear. In this case there is generally a problem with the ossicles, and it is frequently a result of repeated infections in the middle ear. This occurs most commonly in younger people.

Nerve deafness occurs when the nerves fail to transmit signals to the brain, so it is a problem of the inner ear. In general it occurs because of a deterioration of the hair cells or the nerves leading to the brain. It is much more serious than conduction deafness, as little can be done at the present time to restore hair cells or nerves.

The two types can be distinguished by holding a vibrating tuning fork to the head. A person with conduction deafness still has a working inner ear, and the vibrations will be conducted to it by bone conduction. Bone conduction occurs because the skull picks up the vibrations, and they, in turn, set the perilymph in motion, causing a wave within the cochlea that is detected by the hair cells.

One form of nerve deafness is called *presbycusis.* It is the progressive loss of hearing at high frequencies as you age. A certain amount is normal, but the process is hastened if you are exposed to excessive noise. Earlier I mentioned that the usual hearing range for people is 20 to about 15,000 Hz, but older people do not hear frequencies any-

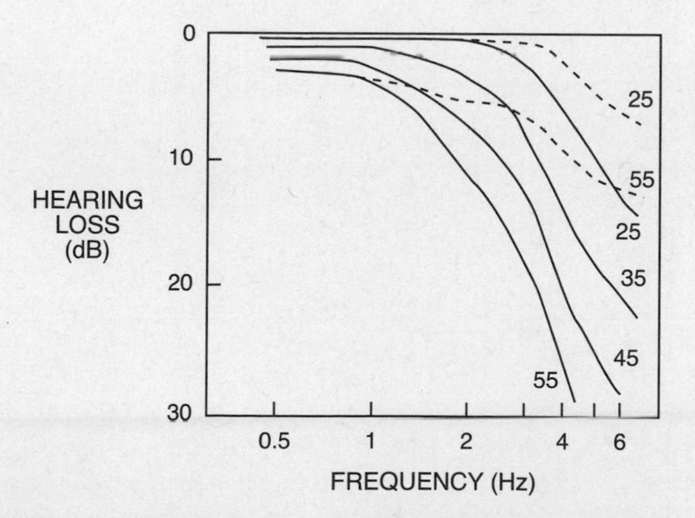

Fig. 21. A plot of hearing loss versus frequency for various ages. Solid curves are for men and dotted curves for women.

where near this high. I used to do an experiment in class using a frequency generator. I would tell the students to put their hands in the air. Then I would gradually increase the frequency of the signal and tell them to lower their hands when they could no longer hear it. And as I was older than most of the students in the class, most hands were still in the air when I could no longer hear anything.

One of the reasons we lose hearing as we get older is a muscle called the tensor tympanic muscle that is attached to the inside of the eardrum to help keep it conical. As we age, this muscle weakens, so that the coupling between the eardrum and the oval window becomes less efficient. This causes a decrease in sensitivity of the ear at high frequencies (fig. 21). Men generally show a greater loss than women at the same age. But you don't need to worry if you are older. Most of the loss is above 5,000 Hz and this doesn't seriously affect your ability to appreciate music, since music rarely goes above 5,000 Hz.

3

Good Vibes

Waves in Motion

I was once seated behind a large pillar at a concert. Needless to say, I was annoyed. The view of the performers was partially blocked, but what I was really worried about was the sound. Would it be distorted by the pillar? This is something we will look into in this chapter. In the previous two chapters we looked at many of the properties of sound and the music that arises from it. In particular, we saw that sound is a longitudinal wave with a particular wavelength and frequency; we also saw how the loudness of sound is measured. In this chapter we will look at several other properties of sound, including reflection, transmission, diffraction, and interference. If these are new terms to you, don't worry; I'll define each of them as I discuss them.

Some of the most important contributions to our understanding of waves were made by the Dutch physicist Christian Huygens. His interest was in light waves rather than sound waves, but his results apply equally well to sound. Newton had put forward the idea that light was composed of particles that he called corpuscles. Huygens was not convinced that Newton's theory was correct. He hypothesized that light was a wave, but there was little interest in his theory at first; it appeared to contradict nature. Water and sound waves appeared to bend around objects, but there was no indication that this was the case with light waves. Although Huygens spent the latter part of his life trying to prove that light was a wave, he did not succeed. It took another 100 years for the proof to come, and it came in the form of several ingenious experiments by the English physicist Thomas Young. Despite his lack of success, Huygens left us with a principle that has important implications for all types of waves, including sound. Now referred to as *Huygens' principle*, it states that each point on a wave front can act as a new source of waves. In other words, wavelets can be generated from any point along the wave. We will see the importance of this principle later in the chapter.

Good Behavior: Waves at a Boundary

Let's begin by considering how waves behave when they strike a boundary. If we know how waves behave, we will know how sound behaves, since sound is a wave. For simplicity we'll start with a single pulse; as we saw earlier, a wave is nothing more than a series of pulses that come one after the other. As in chapter 2 we'll consider a rope that is attached to a rigid post. We'll begin by pulling the rope tight, then jerking it up and down to produce a pulse; this pulse will travel down the rope to the post. What we are interested in is what happens when it gets to the post.

Assume that we watch the pulse travel down the rope. When it strikes the boundary, we see that it is reflected, and a pulse comes back toward us along the rope. Looking at it, we see it is different: it is inverted. The pulse that moved down the rope was a crest, but the one returning is a trough (fig. 22a). In the same way, if you caused a trough to move down the rope, it would reflect as a crest. This can be understood by looking in detail at what happens at the boundary, but first we have to introduce *Newton's third law of motion*. It tells us that for every action there is a reaction; in other words, if you push on something, it pushes back on you. When you hold a garden hose, for example, you feel a reaction force on your hands as water comes out of it.

In the case of the pulse hitting the boundary, the pulse transfers its energy to the post, but the post and connection are rigid and as a result most of the energy is reflected. The action on the post creates a reaction back on the rope which is opposite to the action. As a result the crest is reflected as a trough. The reflected wave has the same wavelength as the incident wave, and the same speed; furthermore, the amplitude of the two waves is nearly the same. (The amplitude

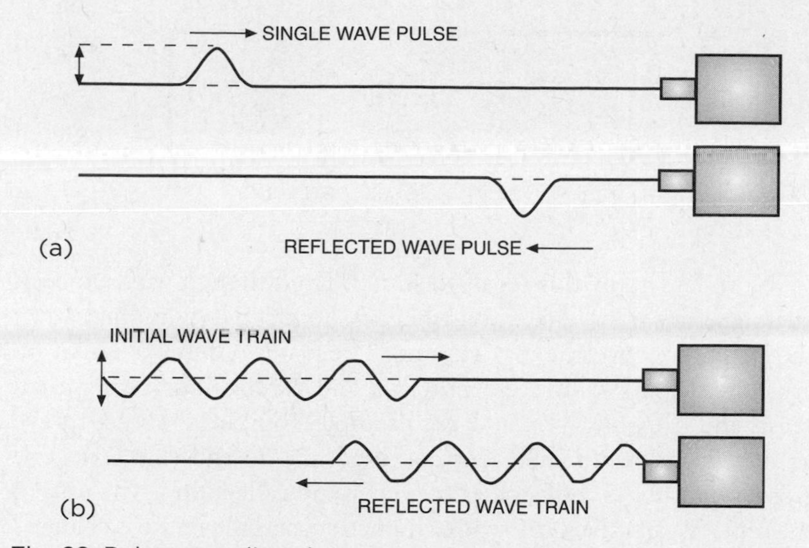

Fig. 22. Pulses traveling down a rope. *Top:* A pulse traveling down a rope that is fixed at its end point. The reflected wave is inverted. *Bottom:* Several pulses traveling down a rope. Reflected waves are also shown.

of the reflected wave may be slightly less, as some energy is lost to the post at the reflection, but in general it is only a small fraction of the total and we'll ignore it.)

We can, of course, extend this to two pulses and more. In fact, if we continue moving the rope up and down periodically, we will create a uniform series of pulses, or a wave, as shown in figure 22b. In practice, the waves that are reflected from the post will interact with the incident waves, but we will leave the details of that until later.

Another case of interest is one in which the rope is loosely connected to the post. The best way to illustrate this *loose end* case is to use a loop to attach the rope to the post so that the connection is not rigid. Again, assume we create a pulse by moving the rope up and down and watch as the pulse is reflected from the other end. Because the connection is not rigid, the action-reaction law that we applied in the previous case does not apply. Instead of an inverted pulse, we get a reflected pulse that has the same properties as the incident pulse. In other words, if a crest was incident on the post, the reflected pulse is also a crest.

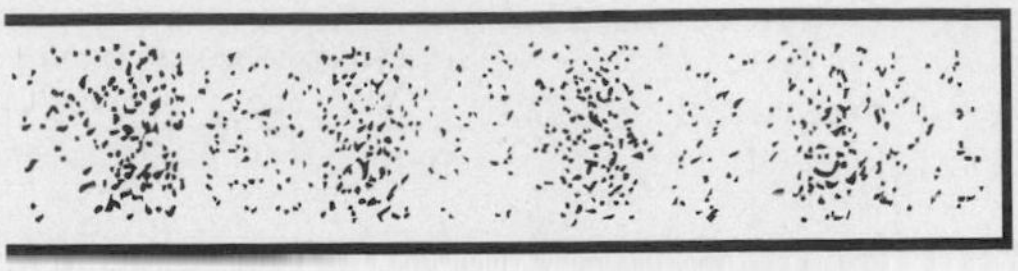

Fig. 23. Longitudinal waves in a closed-end tube.

Now, let's apply this result to sound. Even though we have been dealing with a transverse wave, whereas sound is a longitudinal wave, the results are nevertheless the same. The major difference is that we are now dealing with condensations and rarefactions rather than crests and troughs. We could, for example, think of the sound wave as traveling down a tube that is closed at one end (fig. 23). The closed-end tube is the same as the fixed-end reflection in the case of the rope. We get the same type of reflection, and the phase is changed so that a condensation is reflected as a rarefaction. If, on the other hand, the tube has an open end, we have a situation similar to the free end in the transverse case, so that the reflected wave is the same (it is not inverted).

Entering the Unknown: Waves in a Different Medium

Now let's consider the transmission of a wave into a different medium. Transmission from one medium to another is not as important in relation to sound as the other effects we will be discussing, but it does play a role in acoustics, as we will see later. Again the best approach is to consider a single pulse moving along a rope, but this time the original rope will be attached to another, much thicker rope; we can consider this to be a pulse traveling from a less dense medium to a denser one.

Again, we produce the pulse by jiggling the rope, and watch it as it moves down the rope. As it strikes the thicker rope (or the denser medium), part of it is reflected, and part of it is transmitted to the thicker rope. The reflected pulse is inverted as it was in the case of the rigid connection, but the transmitted pulse is not: it has the same phase as the incident pulse (fig. 24). The transmitted pulse also has a smaller amplitude and wavelength than the incident pulse. Furthermore, since it is in a denser medium, it travels more slowly. The speed and wavelength of the reflected pulse, on the other hand, are the same as those of the incident pulse. A similar phenomenon occurs when the

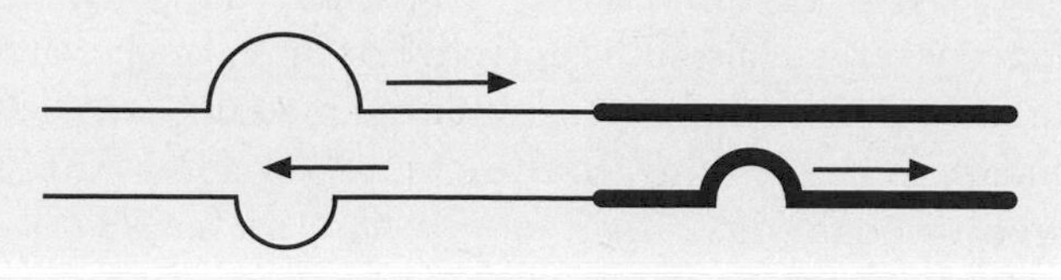

Fig. 24. A pulse moving from a less-dense (thinner) rope to a denser (thicker) one.

first rope is thicker than the second one, but in this case the reflected pulse has the same phase as the incident one (a crest is reflected as a crest), the transmitted pulse has a longer wavelength than the reflected one, and it travels at a higher speed than the reflected pulse.

This result is important in relation to *acoustic impedance*. We talked about impedance matching earlier and noted that it is important when waves are transmitted from one medium to another; in particular, the impedance of the two media should match as closely as possible. If a light rope is connected to a heavy one, as in figure 25a, it is the same as a small impedance connected to a large one. In this case much of the energy of the wave is reflected, and only a small fraction gets through. Furthermore, the amplitude and wavelength of the wave transmitted is changed. The wave in a thick rope approaching a less thick one will produce similar problems; much of the energy of

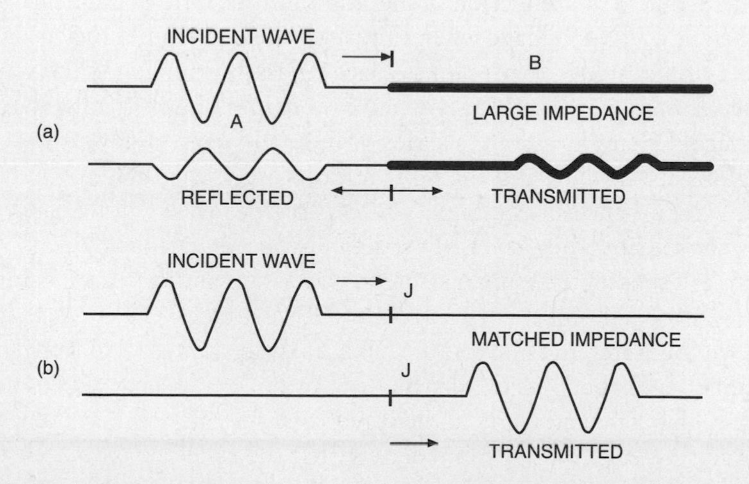

Fig. 25. A simple illustration of impedance matching. (a) Wave interacting with a large impedance. (b) Wave interacting with matched impedance.

the reflected wave will also be reflected. For impedance matching we need the size of the rope (or density of the medium) on each side of the boundary to be the same. In this case all, or at least most, of the energy gets through (fig. 25b).

Impedance matching is important in many situations in music, particularly in relation to musical instruments. Consider a violin, for example. When the strings are vibrated, little noise would be heard without the body of the violin. Because of their small area, the strings do not cause very much air to vibrate. They have to be stretched through a bridge to a large wooden surface, where the impedance match to air is better. In short, the strings set the body of the violin vibrating, and the vibrating body, in turn, causes vibration in a relatively large amount of air. This also applies to the piano. Without the sounding board, we would barely be able to hear the vibrations of the strings. The device for creating impedance is usually referred to as an impedance-matching transformer.

Echo, Echo: Reflection of Waves

All of us have been in a canyon or large empty room or building where we have heard an echo of our own voice. How many times have you tried to count the number of echoes you could hear as you yelled into a canyon. I certainly have—many times. This echo is, of course, due to the reflection of the sound waves. One of the best ways to see how waves behave when they are reflected from a surface is to use a ripple tank. A ripple tank is a large glass-bottomed tank of water that allows you to see waves when you shine a light on the water's surface. What you see are a series of light and dark regions; the dark regions are troughs and the light regions are crests. When you create a disturbance in the tank, you can see the waves move, and it's easy to see how they react when they encounter various objects.

Let's start by creating a straight wave; we can do this by using a straight-edged object such as a ruler. If we attach it to an oscillator above the water and allow it to strike the water, waves will move out from it, and the crests and troughs can be seen clearly. If we now place a barrier at an angle to the wavefront, we can watch as the wave reflects from it. We see immediately that the reflected wave is at the same angle to the normal (or perpendicular) as the incident waves are. This is a result of the *law of reflection* for waves, and of course, it applies to sound waves. The law of reflection tells us that the incident

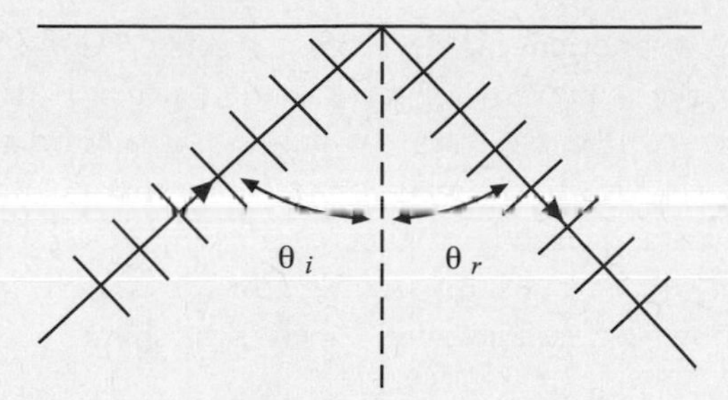

Fig. 26. Reflection of a wave from a surface. The angle of incidence, θ_i, is equal to the angle of reflection, θ_r.

angle (θ_i) will always be equal to the reflected angle (θ_r) when a wave is reflected from a barrier (fig. 26).

Sound waves are not always reflected from flat surfaces, however; sometimes they are reflected from curved surfaces. Consider the inner surface of a sphere (as seen in fig. 27); how would a plane wave be reflected from it? Again, we can check this using a ripple tank. In this case we see that the reflected waves are interacting with one another. Because of the confusion they appear to be causing, it's best to look at the "rays" associated with them rather than the wave fronts. The rays are shown in the diagram as arrows that are perpendicular to the wave fronts. It's easy to see that these rays are all headed for the same general region, but the individual rays do not cross at the same point. You may be familiar with the phenomenon in relation to light; a spherical mirror does not reflect light to a point (fig. 27). In

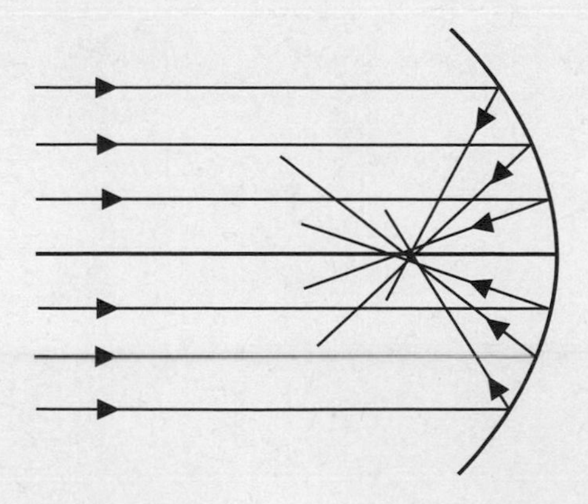

Fig. 27. A wave reflected from a spherically curved surface.

the same way, a spherical surface does not reflect sound to a point; a parabolic surface is needed for this. If the surface is parabolic, the reflected waves will move toward a single point known as the *focal point*. This gives rise to an interesting phenomenon in relation to sound. If you stand exactly at thc focal point of a large parabolic reflector, you will hear distant sounds amazingly well; move off the focal point, however, and you will barely hear them.

I Can't Believe We're Hearing Them: Refraction of Waves

Another important phenomenon in relation to sound waves is refraction. Basically, refraction is the "bending" of a sound wave. If you've ever been out on a lake in a boat during the evening, you've probably encountered it. You can hear people talking on shore, even if they're a long distance away. You know that you would never be able to hear them during the day, so why can you hear them now? The reason is that in the evening, the warm air from near the surface of the lake rises, so you have a layer of warm air above the lake, but the air near the surface is cold. As the sound waves from voices on shore expand outward, they rise to the hotter upper layers, but sound travels faster in warmer air than in cold. This speedup causes the waves to bend, and they are bent toward you, allowing you to hear the sound much better (fig. 28). During the day the warm air is closer to

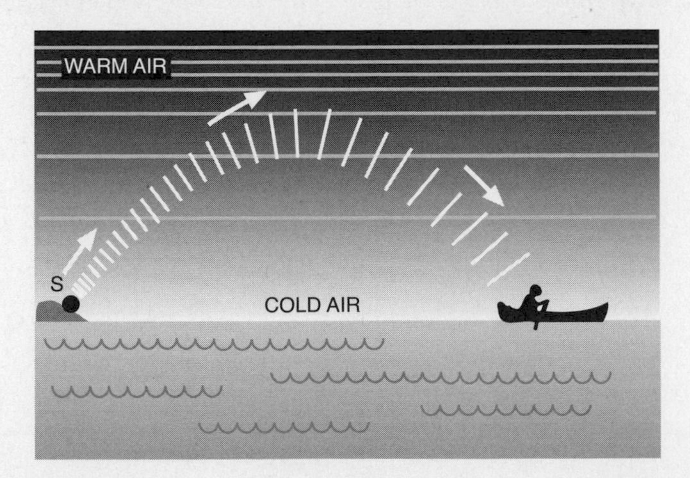

Fig. 28. Refraction of a sound wave from layers in the atmosphere.

the surface and the colder air is up high; this causes a bending in the opposite direction, away from you.

So if you have a situation where sound waves are traveling through two different media, and they travel faster in one of the media than in the other, the wave train will bend. This occurs most frequently when there are different temperatures. The same phenomenon occurs with light; when a light beam enters glass, for example, it travels more slowly and therefore bends, or refracts.

Hey! How Did You Do That? Diffraction of Waves

Waves can also bend around a barrier or a corner. While the idea that waves could bend around a barrier might seem strange at first, it's something you encounter all the time. When someone calls you from another room, you can hear that person if a door is open, even though you can't see the person. This is because the sound wave travels through the door and bends around it. (Actually, some of the sound is also due to reflection from the walls in your room.)

The ripple tank is again one of the best ways to see diffraction. This time the barrier will have a small hole in it. The size of the hole should be approximately the size of the wavelength of the waves (or slightly smaller), and the waves should be plane waves. We see that when the waves strike the barrier, a wave appears to emanate from the hole. In effect, a wave expands out in all directions from it, as in the diagram. There is a wave behind the barrier, next to the hole, so the wave is, indeed, bending around the corners of the hole (fig. 29).

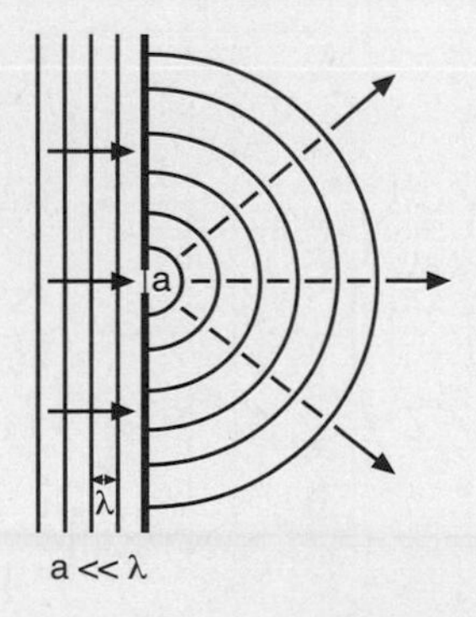

Fig. 29. Diffraction of a wave as it passes through a small opening. The diameter of the opening (a) is much less than the wavelength (λ).

This is diffraction, and it is a result of Huygens's principle. The same phenomenon occurs with light, but the hole has to be much smaller because the wavelength of light is much shorter than that of sound.

This explains why you could hear the voice in the next room through the door. If you vary the frequency of the sound, however, you will notice that some frequencies are heard better than others. If the wavelength is larger than the opening, the amount of diffraction will be larger and you will be able to hear the sound better. Consider an opening of 0.8 m; the frequency corresponding to this wavelength is $f = v/\lambda = 344/0.8 = 430$ Hz, which is approximately that of the note A in the octave of middle C. This mean that notes of a wavelength greater than A (or frequencies less than A) should bend easily around the opening. Shorter wavelengths will not diffract as well and will therefore not be heard as well.

In summary, the theory of diffraction shows that if the size of the opening, a, is smaller than the wavelength of the sound, the amount of diffraction will be large, and the sound will spread out over most of the area behind the opening. If a is approximately equal to the wavelength, there will still be considerable diffraction, but the maximum intensity will be in the forward direction (fig. 30). If a is very much larger than the wavelength, there will be almost no diffraction.

Diffraction also occurs around an object in the path of the sound, and again, the amount depends on the size of the object compared with the wavelength of the sound. In general, as we saw earlier, diffraction is more pronounced for longer wavelengths ($a << \lambda$, where a is the size of the opening). This means you can hear low frequen-

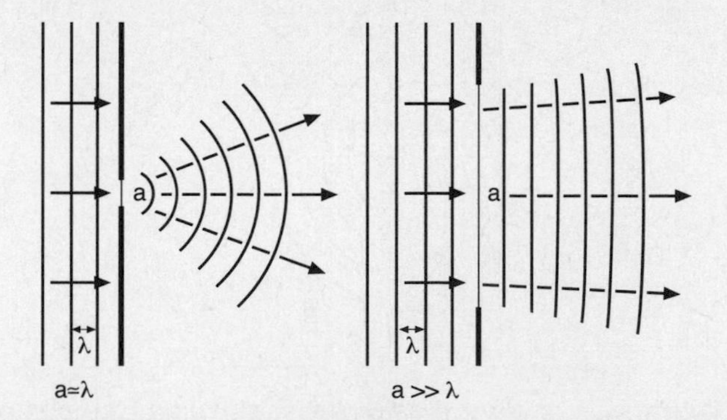

Fig. 30. Diffraction of waves through larger openings.

cies around obstacles better than high frequencies. The pillar that was in front of me at the concert would therefore probably have had little effect for most frequencies. This aspect of diffraction is important in trying to soundproof a room. Small openings will allow sound from the outside to spread easily throughout the room, and they are therefore something you want to avoid.

Another way to look at sound diffraction by an object is to compare it with light. It's a well-known property of light that you cannot see an object that is smaller than the light you are using to illuminate it. Thus, a sound wave cannot "see" an object that has a size much smaller than its wavelength.

Mixing Magic: Interference of Waves

What do you get when you superimpose waves? The easiest way to find out is to go back to our rope. Assume we have two pulses traveling along the rope in opposite directions; assume also that they have the same amplitude and both are crests. When the two crests meet they will interfere with one another. In the case of two crests they will interfere *constructively;* this means that if they both initially have an amplitude of one, when they pass through one another, they will generate a crest of amplitude two (when they are exactly in phase). The same thing occurs when two troughs come together. The new, larger crest (or trough) does not last long, however; the waves pass through one another and continue in the direction they were originally traveling with the same velocity and wavelength they had before the encounter.

Now consider two pulses of the same amplitude that are approaching one another, but with one being a crest and the other a trough. As they pass through one another, the two pulses destroy one another; this is referred to as *destructive interference.* When the amplitudes match, there is no movement of the rope (no pulse), but as they pass this point they continue on as they were.

The two pulses don't have to have equal amplitude for either constructive or destructive interference. If we had a crest of two units meeting a trough of one unit, for example, the interference would create a crest of one unit. Basically what we are doing in determining what the resultant pulse looks like is applying the *principle of superposition.* In the case of waves this principle can be stated as follows:

When two waves interfere, the resulting displacement at any location is the algebraic sum of the displacements of the individual waves at that position.

We can apply this principle at any time while the two waves are interacting. Assume we have a wave with an amplitude of three units that is traveling to the right and one with an amplitude of two units that is traveling to the left. When the two waves come together, we can determine the shape of the resultant wave by applying the principle of superposition at each point along the wave. When the two waves are exactly in phase, we will have the situation shown in figure 31. Applying the principle of superposition we see that the amplitude of the resultant wave will be five units. To get the exact shape of the new wave we would have to apply superposition at each point along the wave.

When applying interference and the principle of superposition to sound we are, of course, dealing with condensations and rarefactions, but the principle is the same. Constructive and destructive interference also occur, and the principle of superposition is applied in the same way. When, for example, two condensations with a unit of one come together, they create a condensation with a unit of two, and so on. Similarly, a condensation and a rarefaction of the same size will cancel one another, and there will be no sound.

The above discussion can easily be extended to two dimensions, and one of the best ways to represent it is with a ripple tank. If a sharp object disturbs the water, waves will spread out in ever-widening circles around the point of disturbance, and the crests and troughs will be easy to see. But if two sharp objects disturb the water at the same time close to one another, a set of rings will develop around each and interfere with one another as they move outward. At some points we will see constructive interference, and at others, destructive interference. We will get a pattern like the one shown in figure 32.

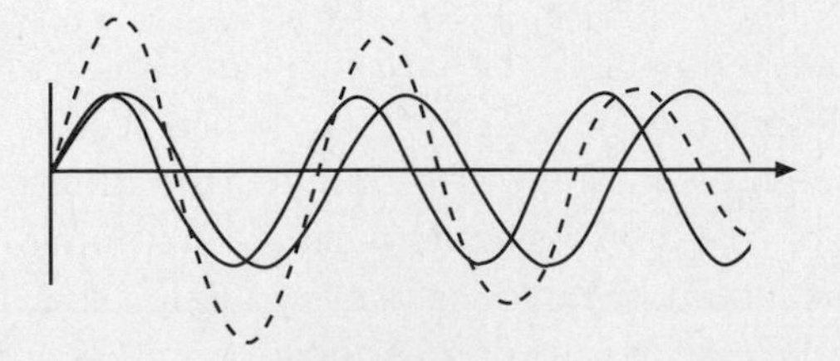

Fig. 31. The superposition of two waves.

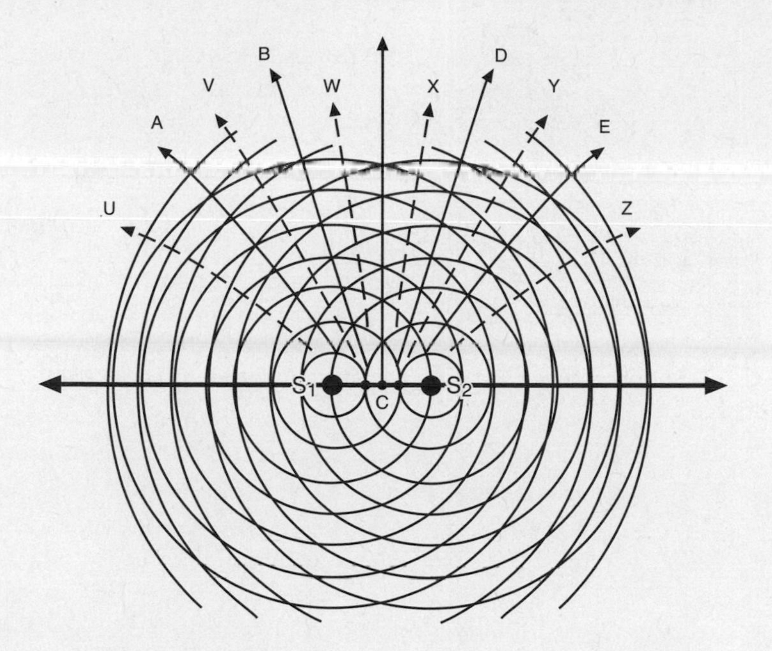

Fig. 32. A ripple tank pattern from two closely spaced disturbances. Dashed lines represent waves that are out of phase, and solid lines, waves in phase.

An excellent way to produce the same effect for sound is to use two loudspeakers, S_1 and S_2, mounted in a box a certain distance apart (fig. 33). We can assume that they both give out a signal of the same frequency. The sound waves will spread out in spheres, and as in the case of the ripple tank, they will interfere with one another. At certain points, compressions will superimpose with other compressions, and rarefaction will superimpose with other rarefactions; these will be regions of loudest sound. At the points where compressions superimpose with rarefactions, no sound will be heard. This means that if you were to walk along the "listener arc" as shown in the figure, you would hear a loud sound at some points and no sound at others. (We are assuming, of course, that no sound is reaching you from reflections off the walls of the room.)

As we will see later, in music the superpositions of certain sounds are more pleasing to the ear than others. Consider the superposition of two notes from a musical instrument where one has double the frequency of the other (assuming that the tones are pure). This range

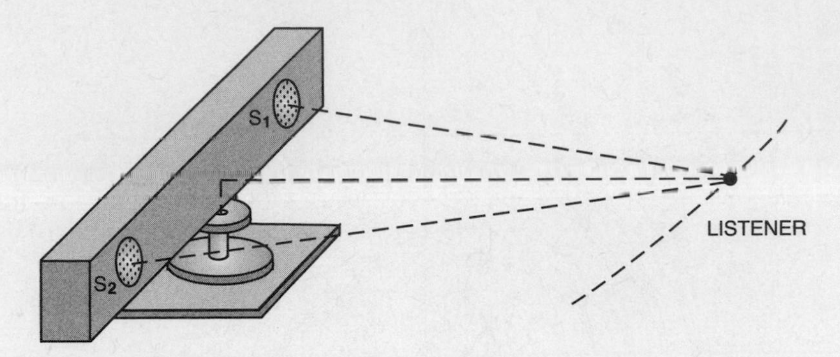

Fig. 33. Sound waves coming to a listener from two separated sources.

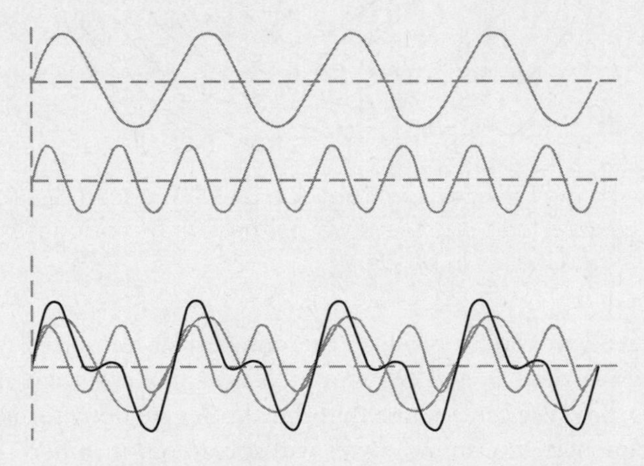

Fig. 34. Superposition of two tones. *Top:* Waveforms of two notes an octave apart; one has twice the frequency of the other. *Bottom:* Combination of the two.

of tones is referred to as an octave, and the result is shown in figure 34. We see that the result is a wave with a regular, repeating pattern. It is pleasing to the ear and can be considered to be musical. In the same way if we superimpose two frequencies that are in the ratio 3:2, we find that they also give a similar pleasing regular pattern. In this case, the two notes are a fifth apart, which is an important interval in music. On the other hand, if we superimpose two frequencies where there is no clear mathematical relationship between the frequencies, the resultant pattern will not be regular, and it will not be pleasing to the ear. We refer to it as "noise."

If two notes of slightly different frequencies are sounded, we hear what are called *beats*. Beats occur, for example, when two musical instruments, such as two violins, sound the same note, but one is slightly out of tune. A good way to demonstrate this is to use two tuning forks of the same frequency. If a rubber band is looped tightly around the prongs of one of the tuning forks, it will change frequency slightly. If the two tuning forks are then sounded at the same time, the sound intensity from them will rise and fall periodically. In short, beats will be produced. Because the two waves are slightly different in frequency, they will arrive at your ear out of phase, then in phase moments later, then out of phase, and so on (fig. 35). The beat frequency will be the difference of the two frequencies, $f_{beat} = f_1 - f_2$, where f_1 and f_2 are the two frequencies. The overall frequency of the two fused tones will be $f = (f_1 + f_2)/2$.

Beats are used by piano tuners in tuning a piano. The tuner sounds a note on the piano, then taps a tuning fork of the same frequency. If beats are detectable, the frequency of the note is slightly off. The tuner then tightens or loosens the string until no beats are heard.

Motionless Sound: Standing Waves

We saw earlier that when a wave is reflected from a fixed barrier, it interferes with incident waves, and the two superimpose. In most cases the incident and reflected waves superimpose to give an irregular and nonrepeating wave. In fact, the wave pattern would proba-

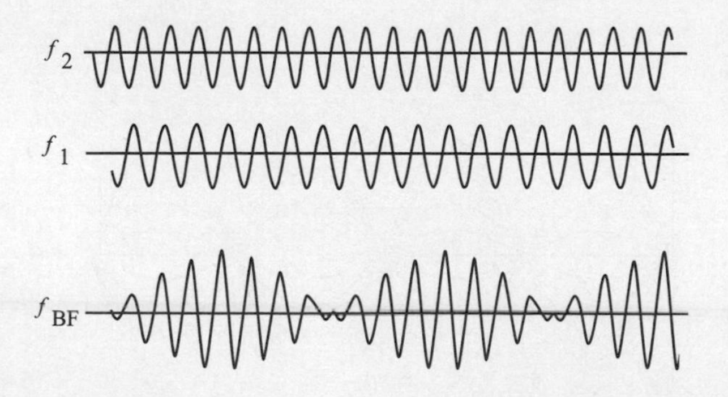

Fig. 35. The creation of "beats" from two closely associated frequencies (BF = beat frequency).

bly be difficult to discern in the midst of all the irregular motions. You can, however, produce a regular wave pattern; to see how, consider a rope that is attached at one end again. If you give the rope exactly the right frequency and do it in the right way, you can get a *standing wave*. In this case, points along the wave will appear to be standing still. They are referred to as *nodes*. In the regions between the nodes, the displacement of the rope will change, but in a regular manner; it will vibrate back and forth.

You need perfect timing to create such a wave; you have to introduce a wave crest on the rope exactly when a wave crest begins its reflection from the far end. In this case there will be complete waves moving to the right and to the left at all times, and interference will occur when the right-moving waves meet the left-moving waves. The waves will interfere both constructively and destructively; when they interfere destructively, they will leave periodic points of no displacement along the rope. These are the nodes, which are usually designated by *N*. When they interfere constructively, they leave regions of large displacement; these regions, which are between the nodes, are called *antinodes* (*AN*), shown in figure 36.

A standing wave is not actually a wave; it is a pattern that results from the interference of two waves of the same frequency and amplitude, with different directions of travel. The nodes and antinodes are a result of the interference of the two waves. Nodes are formed where the crests of one of the waves meets a trough of the second wave, or a half crest of one wave meets a half crest of the other, and so on. Antinodes are produced at locations where constructive interference occurs—for example, when a crest of one wave meets that of a second wave.

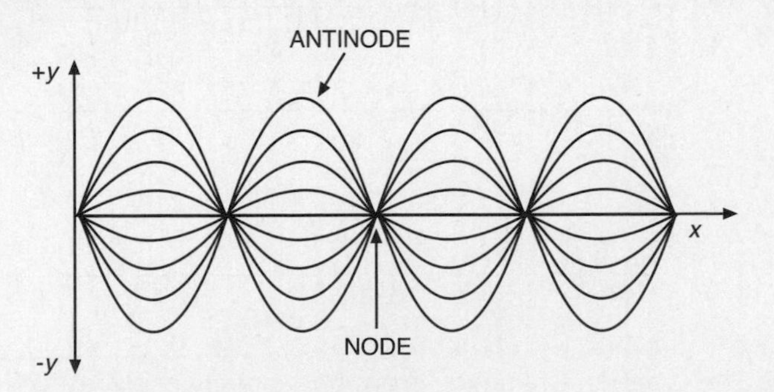

Fig. 36. A standing wave, showing nodes (N) and antinodes (AN).

Antinodes are always in motion, vibrating back and forth between large positive and negative displacements. It can easily be seen from the diagram that the length of each loop of the standing wave (L) is half a wavelength—in mathematical terms, $L = \lambda/2$. As we will see later, standing waves are important in musical instruments.

The Doppler Effect

Another interesting effect associated with sound waves is the Doppler effect, named after the Austrian physicist Christian Doppler. He was the first to explain the effect correctly and give a formula for it. You are no doubt familiar with the phenomenon in relation to an approaching car or train. If the car has its horn blaring, you hear a distinct change in its pitch as the car passes you. Doppler explained that the sound waves from a source that is moving toward you will be squeezed together and will therefore have a shorter wavelength (and higher frequency) than they had at rest. When the source is moving away from you, the waves get stretched out and the pitch is lowered (fig. 37).

Doppler worked out the mathematical relation for the phenomenon in 1842. In particular, he showed that the effect occurred both

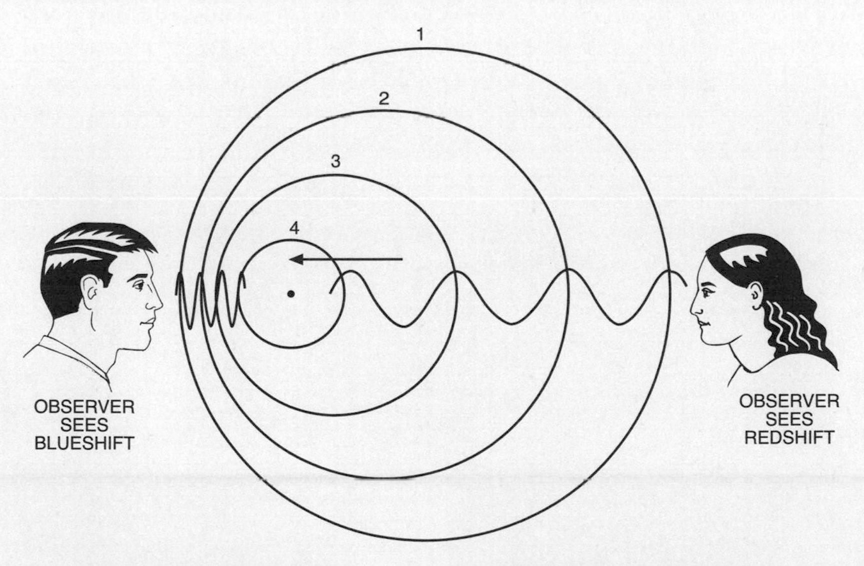

Fig. 37. A simple illustration of the Doppler effect. The case for light is shown, but the effect is the same for sound.

when the source was moving (and the observer was stationary) and when the observer was moving (and the source was stationary). A few years after Doppler explained the effect, it was demonstrated experimentally using a train pulling a flatcar. Trumpeters were stationed on the flatcar, and musicians beside the tracks noted the pitch as the train passed. Doppler's formulas were verified.

THE BUILDING BLOCKS OF MUSIC

II

4

Making Music Beautiful

Complex Musical Tones

Have you ever asked yourself while listening to a recording, or while at a concert, what it is that makes the music so appealing? There are, of course, many answers. The music may bring back memories, or make you feel warm all over, or let you drift off into space (at least it seems that way), or it may just be that it's pleasing to your ear. Regardless of what conclusion you come to, there's no doubt that music makes you feel good. So we have to ask, What is it that does this? Melody is certainly important, as is rhythm, but there's something else you might not realize. The "richness" and complexity of the harmonics also add to the enjoyment. When you play a single note on a piano, or any other instrument, you may think you're getting a single tone, or single frequency, but you aren't. Many different frequencies are striking your eardrum, and of course, the music coming from an orchestra is much more complex. In this chapter we will look at this complexity and explore why it helps make music more appealing.

"Seeing" Music

If you listen to a note—say middle C—played on several different instruments, such as a violin, piano, and a clarinet, you can easily tell which instrument it came from. All are vibrating with the same frequency, 256 Hz, but even if they all have the same loudness, they still sound different. And we can easily hear the difference. But if they are different, there has to be a way to see this difference. And, indeed, there is. What we need is a way of looking directly at the sound so that we can actually "see" the music. And for this we need two devices: a microphone and an oscilloscope. Both are relatively complex, so I won't go into the details of how they work, but I will give a brief outline. Later in the book I'll talk about them in more detail.

The major part of a microphone is a pair of electrically charged metal plates. The outer one, which is referred to as the diaphragm, is thin enough so that it vibrates when an air pressure wave such as the one created by your voice strikes it. These vibrations cause tiny electrical currents to flow in an external circuit that are proportional to the amplitude of the diaphragm; the current is therefore a "coded" copy of the oscillations. In effect, the vibrating signal has been converted to an equivalent electrical signal.

This oscillating electrical current is then fed to an oscilloscope. If you're not sure what an oscilloscope is, you merely have to look in your living room or den; the heart of your television set is an oscilloscope. In a television set a beam of light sweeps across the screen thousands of times a second. After each sweep it moves down slightly so that it eventually sweeps the entire screen. This beam causes the screen to glow with a particular intensity, and since the intensity at each point is continually changing, we see a picture.

In the same way, the oscillating current from our microphone is fed to two metal plates in an oscilloscope (fig. 38). A beam passes through the region between these plates and is deflected according to the charge on the plates; in other words, it changes in the same way that the oscillating electrical current that is applied to it changes. Finally, as in the case of the television set, the beam is moved rapidly across the screen hundreds of times per second until it has covered the complete screen.

What we see is a "picture" of the sound wave that struck the microphone. If the sound is pure, such as that from a tuning fork, we get a perfect sine wave and can easily determine its frequency and wavelength by measurements made on the screen. But when the note from a musical instrument is projected on the screen, we see immediately that it looks quite different. And we can now answer the question: why does a note such as middle C sound different on a violin, a

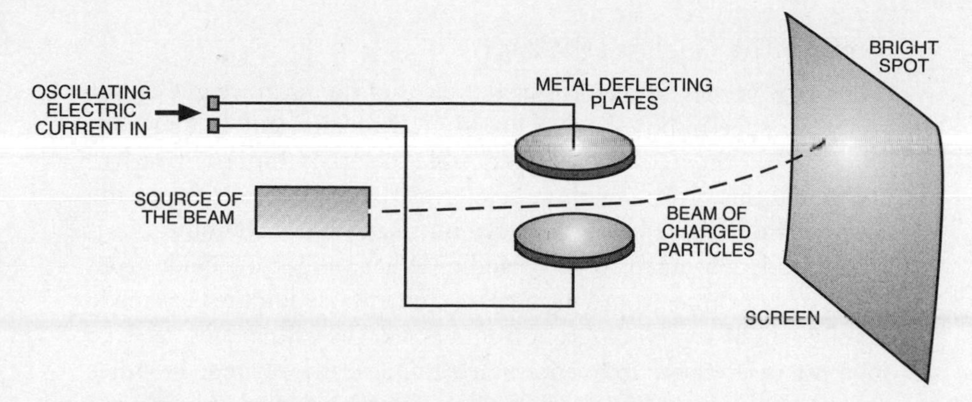

Fig. 38. A simple representation of an oscilloscope.

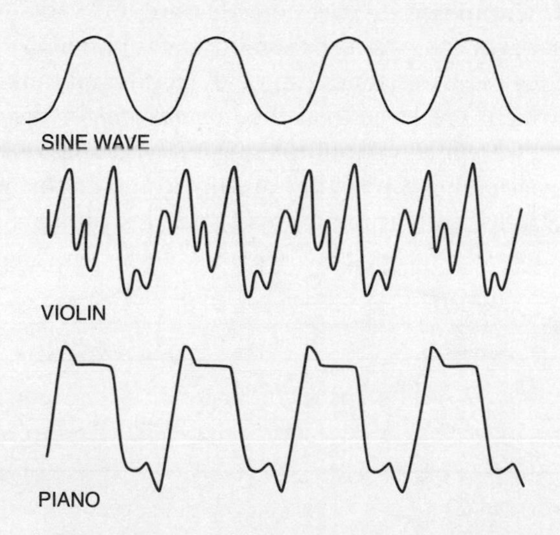

Fig. 39. The same note sounded on a signal generator, a violin, and a piano.

piano, and a clarinet. If we look at the sound from each of the instruments, we see that each has a frequency of 256 Hz, as expected, and each has the same loudness (with the height, or amplitude, of the wave measuring the loudness), but other than that their waveforms are quite different (fig. 39).

Timbre: The Quality of Music

The shape of these waveforms reminds us of the form we got earlier when we superimposed two pure signals of different frequencies, where one of the frequencies was double the other. Indeed, if we had continued this process with wavelengths that were multiples of the first, we would have made the wave more and more complex, but it would have continued to be periodic. What we can conclude from this is that any tone from a musical instrument, say middle C, is made up of waves of several different frequencies. We will, in fact, see that in most cases these frequencies are numerically related; in other words, they are multiples of the first frequency. And this is what makes the same note from various musical instruments different. Each of them has the same overall frequency, but they have other frequencies superimposed on this note that are different. We refer to these other waves as *overtones* or sometimes as partials.

In practice these overtones can be exact integral multiples of the first tone, which is referred to as the *fundamental*, or they can be arbitrary. If they are integral multiples, as shown in figure 40, we refer to them as harmonic; if not, they are inharmonic. For most instruments, overtones are harmonic; only such instruments as cymbals

Fig. 40. Overtones. The top one is the fundamental.

and bells have inharmonic overtones. So we will direct almost all our attention to harmonic overtones.

The difference in the waveform of a note from instrument to instrument is referred to as the *timbre*, or *quality*, of the tone. Without thinking about it, you encounter timbre every day. The human voice is also made up of various overtones, and because of this, each voice is distinctive. This is why you can identify someone over the phone so easily, even though you can't see the speaker.

While the timbre, or quality, of a tone is mainly a result of overtones, other things also contribute to it. Consider a violin string, for example. You know that it sounds different when you pluck it, compared to when you bow it. We refer to this difference in the quality of sound as being the result of the *attack*, or method of producing the sound. Also important is the *decay* of the note—in other words, how long it takes for the sound to fade away.

Complex Tones: Analyzing the Music

Although a musical note is composed of many different frequencies, it can be broken down into pure tones or single frequencies in a process referred to as *analysis*. This is now relatively easy to do with modern electronic instruments. Also important in music is the converse process, namely the bringing together of many frequencies to produce a complex sound. The combining of frequencies is referred to as *synthesis* and is done by electronic instruments called synthesizers.

Let's look at synthesizers in more detail. A question that immediately comes to mind is whether it is possible to produce a waveform of any shape if we add enough harmonics together? The answer is yes. And the man that proved that it could be done was Jean-Baptiste Fourier of France. There is little indication that Fourier was particularly interested in music, or even sound; his major interest was in how heat flowed from one point to another, and he made many important contributions to the theory of heat. In the process, however, he formulated what is now known as Fourier's theorem; it applies to all waves, and since sound, and music, are waves, it applies to them.

Fourier's theorem can be stated as follows:

> Any periodic oscillation curve, with frequency f, can be broken up, or analyzed, into a set of simple sine curves of frequencies f, $2f$, $3f$, . . . each with its own amplitude.

In the case of sound, these "simple sine curves," or waves, are harmonics; and as we saw, we refer to the first as the fundamental, and the higher multiples of it as overtones. This means, for example, if the fundamental has a frequency of 200 Hz, the first harmonic is 400 Hz, the second is 600 Hz, and so on. All these frequencies are sounded at the same time, so that when a musician plays a single note, he is actually playing several frequencies. Furthermore, if the same note is played by two musicians, say, two violinists, the two notes will not be identical, even if the two violins are perfectly in tune. The reason is that no two instruments are exactly the same structurally; also, no two musicians bow the instrument in the same way. The result will be beat notes between the two violins; in fact, beats will even occur between the second, third, and higher harmonics. This, however, does not detract from the sound; the overall effect is called the chorus effect, and it is something that adds to the richness of the sound.

Harmonic Spectra

One of the best ways to show what overtones are present and what their amplitudes are, is by using a bar graph. It is a plot of frequency versus loudness, or amplitude, but since the tones are distinct, it looks like a series of vertical lines. Frequency is plotted along the horizontal, and relative intensity is plotted in the vertical direction, Usually the fundamental is assigned a value 1.0, and the overtones are compared to it. The plot of a pure tone is shown in figure 41. Most instruments, however, have relatively complex spectra, as seen in figure 42.

These graphs give us an excellent way of "seeing" musical notes. We can see immediately what overtones are present and also their intensities. Not only does the spectrum of different instruments differ,

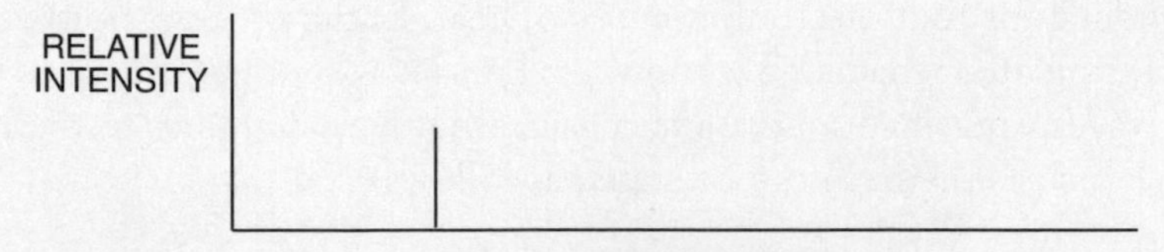

Fig. 41. Bar graph (spectrum) of a pure tone.

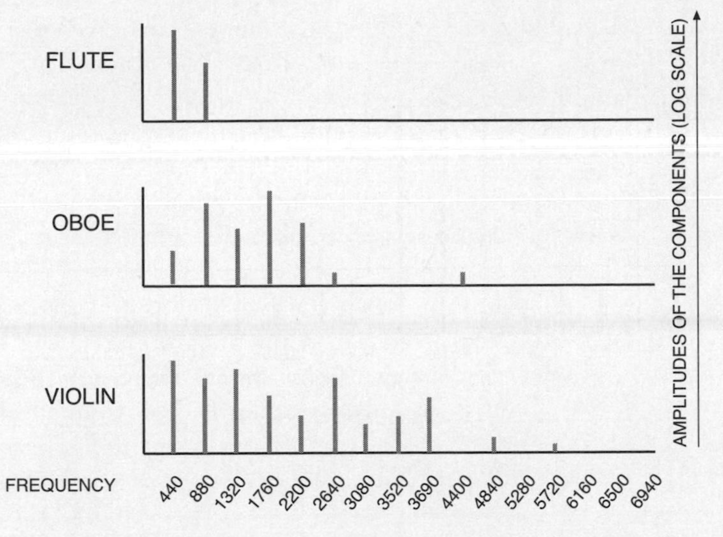

Fig. 42. Bar graph (spectrum of frequencies) of a flute, an oboe, and a violin.

but the spectrum within a single instrument depends on what note is played (the note C, for example, will give a different spectrum than the note F).

Formants

Since each instrument has its own distinct spectrum of harmonics, it might seem that the spectrum for a given type of instrument would always be the same. But this isn't so. Several things beside the spectrum of harmonics characterize an instrument. One of the most important is that the harmonic structure depends on loudness. Loud notes usually contain many more high-frequency harmonics. In addition, the musician playing the note makes a difference; each musician plays it slightly differently. And as we saw earlier, the attack and decay of the note also make a difference. Because of this, it is useful to supplement the harmonic spectrum of an instrument with its *formant.* The formant of a musical note is a frequency region where most of the sound energy is concentrated (fig. 43). It might seem that this region would consist of the frequencies near the fundamental, but this is not necessarily the case. Often, the high harmonics are louder and determine the timbre of the instrument.

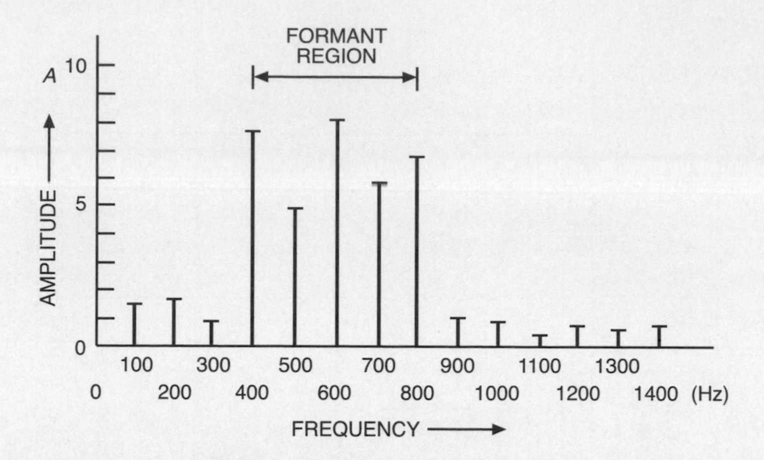

Fig. 43. Bar graph showing the formant region.

How Can It Vibrate That Way? Vibrational Modes of a Stretched String

When a string vibrates with several harmonics, they are all vibrating at the same time. This may seem like a difficult thing for a string to do. How can it vibrate in several ways all at once? Since several instruments, including the violin, the guitar, and the piano, have vibrating strings, it is instructive to look at the various vibrational modes of a stretched string.

Consider a string of a certain length that is vibrating at its resonant or natural frequency. Each natural frequency produces its own characteristic vibrational mode, or standing wave pattern, and it is these standing wave patterns that we will be looking at.

Let's begin by attaching the string at two points, as in the diagram shown in figure 44. The two ends are unable to move and are therefore nodes; in between these two nodes are one or more antinodes. If there is only one antinode, this harmonic is the *fundamental;* it is also sometimes called the first harmonic. This harmonic will have the longest wavelength; in fact, the wavelength will be twice the length between the two nodes (fig. 45).

The second harmonic, or first overtone, is produced when the string vibrates with a node in the center. In this case it will have three nodes and two antinodes (shown in fig. 46). We see from the diagram that exactly one wavelength fits between the two end nodes, so the

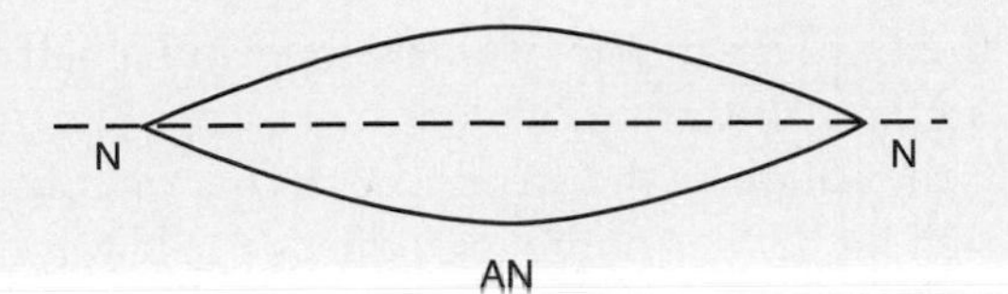

Fig. 44. The fundamental of a string.

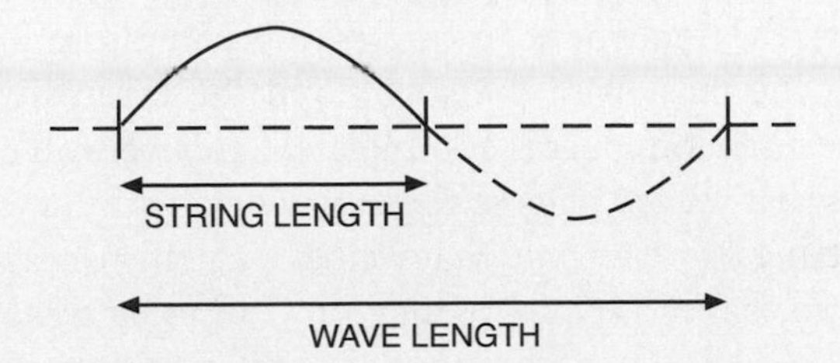

Fig. 45. Note that the wavelength is double the string length.

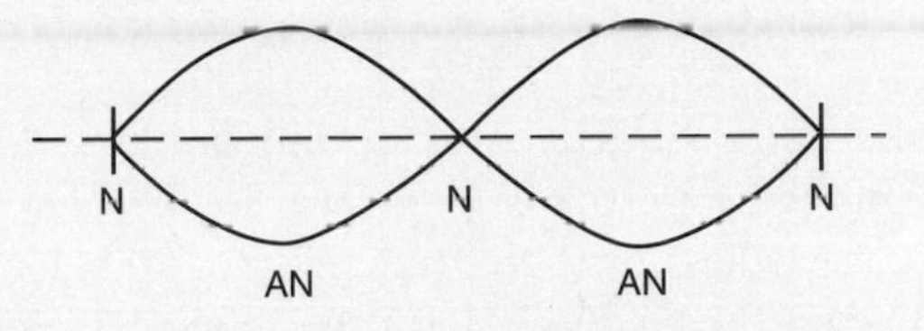

Fig. 46. The first overtone (or second harmonic) has one wavelength, with three nodes and two antinodes.

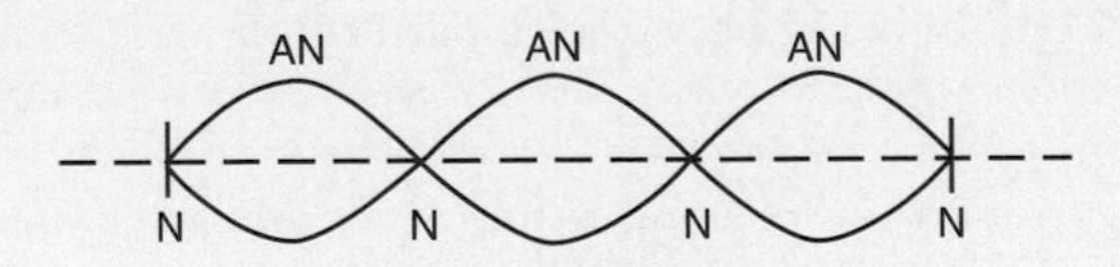

Fig. 47. The second overtone (or third harmonic).

wavelength (λ) is equal to the string length (*L*). For the third harmonic, or second overtone, we have to add another node; in all there will therefore be four nodes and three antinodes (fig. 47).

The length of each loop in the higher harmonics is the same. In the case of the third harmonic, we have one and a half wavelengths along the length of the string, which means there is 3/2 of a wavelength along the length of the string. With this we can see a pattern emerging: each higher harmonic introduces a half-wavelength. This means that the fundamental has $L = \lambda/2$, the first harmonic has $L = 2/2\lambda = \lambda$, and the third harmonic has $L = 3/2\lambda$ and so on, and we can easily solve each of these for λ. Our results are summarized in table 3.

So far we have said nothing about frequencies. It's well-known, however, that the frequency of a string—for example, a guitar string—depends on the tension in the string and the linear density of the string (the expression for it is complicated, and we won't get into it). This means that we can change the frequency by tightening or loosening the string. If we assume that we have a string of, say, 70 cm, we can tune it so that it has a frequency of 375 Hz by tightening it appropriately. We also know that there is a relation between speed (v), frequency (f), and wavelength (λ), namely, $v = \lambda f$. From table 3 we can write $\lambda = 2L/n$, where n is an integer, so we can calculate the speed of the waves:

$$375 \text{ Hz} \times \lambda = 375\ (1.4) = 525 \text{ m/sec.}$$

But the speed of the wave is dependent only on the tension and density, and not on the properties of the wave, so all waves will have the same velocity regardless of their frequency or wavelength. We can therefore calculate the frequency of the second harmonic from $v = \lambda_2 f_2$ where $\lambda_2 = L$.

$$f_2 = v/\lambda_2 = 525/.7 = 750 \text{ Hz.}$$

In the same way we can calculate the frequency of the third harmonic; it is 1,125 Hz. Again we see a pattern; it's easy to see that f_2 is

Table 3. Nodes and antinodes for various harmonics in strings and in tubes

Harmonic	Wavelengths	Strings		Tubes		Length-wavelength rel.
		Nodes	Antinodes	Nodes	Antinodes	
1	1/2	2	1	1	2	$\lambda = 2L$
2	1	3	2	2	3	$\lambda = L$
3	3/2	4	3	3	4	$\lambda = 2/3L$
4	2	5	4	4	5	$\lambda = 1/2L$
5	5/2	6	5	5	6	$\lambda = 2/5L$

Fig. 48. All the harmonics vibrating at the same time.

$2f_1$, f_3 is $3f_1$ and so on; in other words, the upper harmonics are integral multiples of the fundamental, as we would expect.

It's important to remember that all these harmonics are vibrating at the same time, as shown in figure 48.

Listening to Overtones on a Piano

Since the harmonic series is the sequence of frequencies *nf*, where *f* is the fundamental and *n* is an integer, we can express this using musical staff notation. For example, for the harmonics of the key A_4 (the A above middle C) we have

The frequency of each note has been specified; you can see that they are integral multiples of 55 Hz.

You can actually play this series on a piano and hear the various harmonics. Begin by slowly depressing the key A_4 and holding it down. This lifts the damper for the key but does not sound it. Now go up one octave to A_5 and strike it hard (and staccato, or detached). After the sound from A_5 has died away, you will hear A_4 vibrating in its second harmonic. Then do the same thing with E; you will hear A_4 vibrating in its third harmonic. You can continue up the keyboard in this way, striking A, C♯, E, and so on, for the third, fourth, and fifth harmonics.

Vibrational Modes of a Column of Air

We have seen the vibrational modes of strings, which are common to violins, guitars, and pianos, but many musical instruments make their sounds using vibrating columns of air. There are, in fact, open-end air column instruments such as the flute, the trombone, the saxophone, and the oboe, and closed-end air column instruments such as the clarinet and the muted trumpet.

A simple form of each of these instruments is a cylindrical tube. When a disturbance (such as blowing in the end of the tube) is introduced, harmonics are set up, and of course, each harmonic is associated with a different standing wave pattern. When a sound wave (which is an air-pressure wave) travels down the cylindrical tube, it eventually comes to the end. But the end acts as a boundary and therefore has an effect on the wave, depending on whether the end of the tube is open or closed. Reflection, partial reflection, transmission, or partial transmission will occur depending on the type of boundary. We will consider open-air tubes (with two open ends) and closed-end tubes (with one end closed and one open). Inversion of the reflected portion of the wave occurs only if the end is closed, so it is analogous to the string with a fixed end.

In reality, compressions and rarefactions in air pressure occur within the tube, but most of the time we will plot pressure variations, so the wave will look like a transverse wave. It's important to remember, however, that it is actually a pressure wave. Let's begin with a cylinder with two open ends. In an open-end tube, if a compression is introduced into one end of the tube, it will reflect as a compression; in other words, there will not be an inversion. Now, suppose that exactly when the reflection occurs at the far end, we introduce a rarefaction at the near end. This rarefaction will travel down the tube and interfere with the reflected compression; in particular, the two waves will interfere destructively at the center of the tube, and there will be a node. In this case (the fundamental), half a wavelength fills the length of the tube, and since we always have an alternating pattern of nodes and antinodes, the two open ends will have antinodes (fig. 49). Assume that we do this over and over so that we have a standing wave; the pressure at the open ends will therefore alternate between high and low, and the center will be at normal pressure.

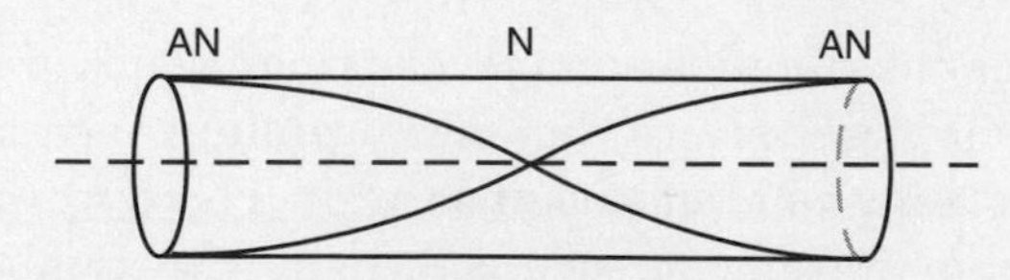

Fig. 49. Fundamental waves in a tube with two open ends.

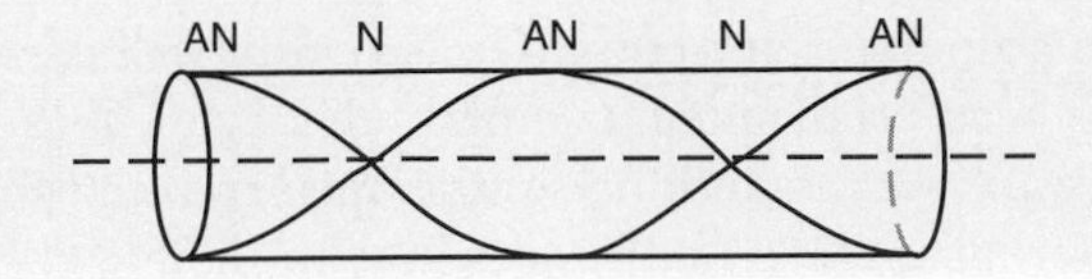

Fig. 50. The second harmonic for a tube with two open ends.

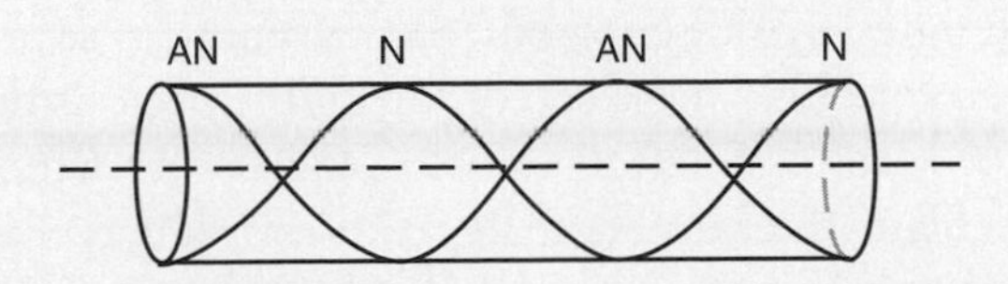

Fig. 51. The third harmonic for a tube with two open ends.

For the second harmonic, as in the case of the string, we add another pressure node. This will give rise to two nodes, and three antinodes (fig. 50). The wavelength in this case is equal to the length of the tube. Similarly, the standing wave pattern for the third harmonic is obtained by adding another pressure node. In this case we have three nodes and four antinodes, and there are one and a half wavelengths along the tube. We can continue in this way to the fourth, fifth and higher harmonics (fig. 51). Table 3 shows the positions of the nodes and antinodes for the various harmonics. The frequency for this case is given by $f_n = v/\lambda_n = nv/2L$.

Closed-End Air Column

In the case we now consider, one end of the tube is closed. A musical instrument of this type is the clarinet. If a sound wave is introduced at the open end of the tube, reflection will occur at the closed end, and inversion will occur. This means that if a compression is introduced at the open end, it will be reflected as a rarefaction, and the rarefaction will move back toward the open end. In the process it will interfere with any waves heading in the opposite direction. If we continue to introduce sound into the tube, we can set up a standing wave.

Suppose that we now introduce a rarefaction into the tube the moment the reflected rarefaction reaches the open end. The returning rarefaction will interfere with it, and since there is one-quarter of a wavelength in the tube, the two waves will constructively interfere to produce a double rarefaction. Constructive interference will, in fact, always occur at the open end of a tube. We will therefore have a standing wave with an antinode at the open end, and since we must have an alternating pattern of nodes and antinodes, the closed end will always be a pressure node. The pressure at the open end will therefore oscillate between high and low pressure, but the closed end will remain at normal pressure. We therefore have the first harmonic as shown in figure 52. The closed end prevents the oscillation of air, and it is therefore like the fixed end of a string. Air at the open end, on the other hand, is free to oscillate in and out of the tube as the pressure oscillates back and forth.

For the first harmonic we see that one-quarter of a wavelength is contained in the length of the tube. For the next harmonic we have to add an extra node; this means an extra antinode is also added, as is depicted in the diagram in figure 53. In this case there is three-quarters of a wavelength in the tube, which is three times the number of waves in the first harmonic. This is therefore the third harmonic. We see that there is no second harmonic; indeed, closed-end tubes such as this possess only odd-numbered harmonics. The next harmonic is

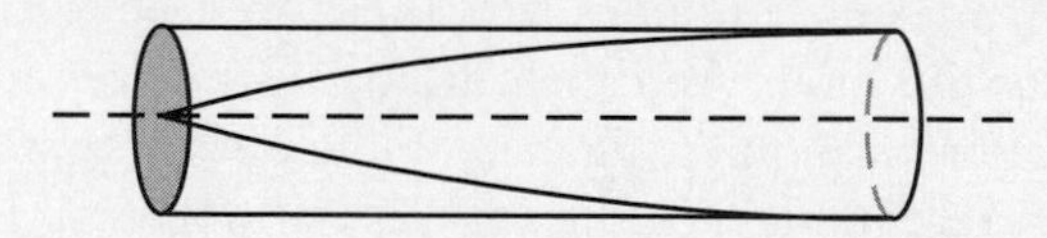

Fig. 52. The fundamental for a tube with one end closed and one open.

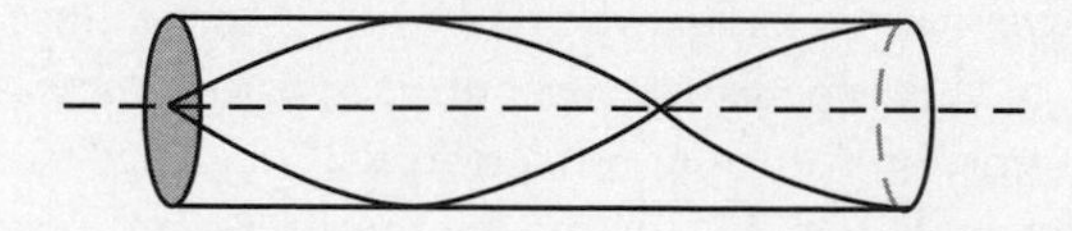

Fig. 53. The first overtone for a tube with one end closed and one open.

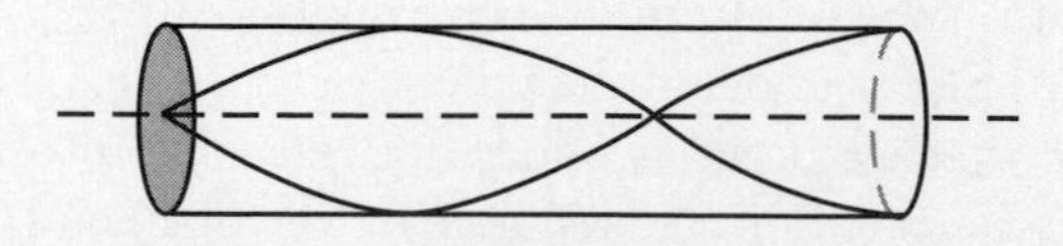

Fig. 54. The second overtone for a tube with one end open and one closed.

therefore the fifth harmonic, and it is shown in figure 54. The equation for the wavelengths of these harmonics is as follows:

$$\lambda_n = 4L/n \qquad n = 1,3,5,7 \ldots$$

Thus, we see that a semiclosed tube can vibrate with many harmonics, but the even harmonics are absent. In addition, the fundamental

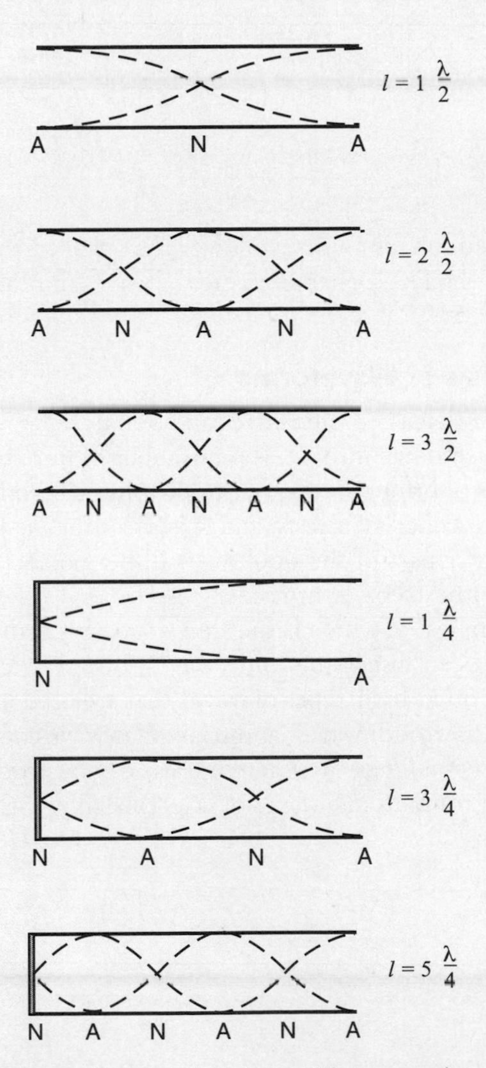

Fig. 55. A summary of waves in tubes with two ends open and one closed and one open.

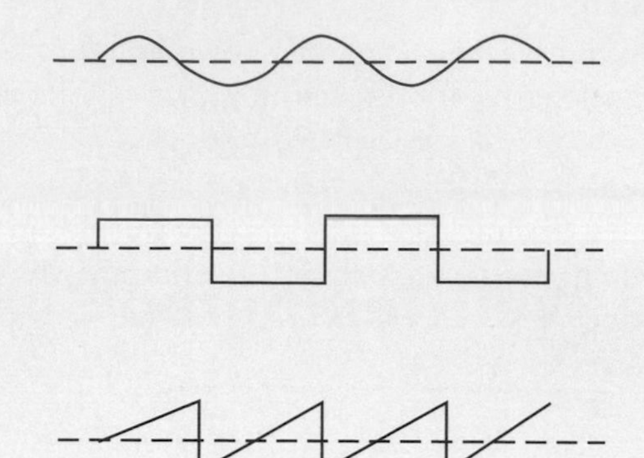

Fig. 56. A sine wave, a square wave, and a sawtooth wave.

of a semiclosed tube is an octave lower than that of an open tube of the same length. These characteristics of waves in tubes are summarized in figure 55.

The Synthesis of Waveforms

We saw earlier that a complex wave can be analyzed—in other words, it can be broken down into a certain number of pure tones. The opposite process of bringing together basic tones to produce a pleasant musical tone is called synthesis, and it is accomplished by synthesizers. In the latter part of the book we will be looking extensively at electronic synthesizers. Synthesizers use several different types of waves, the simplest of which is the sine wave; this is the "pure" wave that we have been discussing. Sine waves, however, can be used to create several other basic types of waves that are used in synthesizers: square waves, sawtooth waves, and triangular waves of various types. They are shown in figure 56. Each of these can be produced by combining the appropriate sine waves or, more usually, with various electronic circuits.

5

The Well-Tempered Scale

In the first chapter we saw that Pythagoras used a monochord—a hollow sounding box with strings stretched across it—to set up a scale. On his box were movable bridge stops that allowed him to separate the string into smaller sections. He found that if he sounded a string of a certain length not stopped with a bridge and another one where a bridge was set one-third of the way along a string of the same length, the overall sound was pleasing to the ear. Although he might not have realized it, these two vibrations are what we refer to as a fifth apart, and this is what he used to build his scale. As we saw, however, his scale had only five notes, and it wasn't long before others were experimenting with different scales. In this chapter we will look at some of these other scales.

Before we continue, however, let's go back to the question of what a scale is and why it is useful. Starting with a definition, we can define a scale simply as a sequence of notes of particular frequencies that are pleasing to the ear. More exactly, it is a sequence of tones whose frequencies are such that their intervals are in the ratio of integral numbers. But why are scales important? The reason is simple: without scales, music would not be possible (at least, music you would want to listen to). They are the basis of all music.

As we will see in this chapter there are many different types of scales. Almost everyone is familiar with the major scales on the piano. It is the sequence of tones: do-re-mi-fa-sol-la-ti-do used by singers. But there are many other types of scales. Minor scales are also used extensively in music, and pentatonic and blues scales are used extensively by jazz musicians and others.

Consonance, Dissonance, and Harmonics

Scales are based on consonance. We say that two tones are consonant if, when they are sounded together, the resulting tone is pleasing to the ear. As we saw, Pythagoras and others found that when the tones formed fifths, fourths, or thirds they were most harmonic, or consonant. But it is easy to find two notes on the piano that are not harmonious. If you sound two notes next to one another, such as C and D, they seem to clash. In other words, the sound they generate is not consonant; we say that it is dissonant. In simple terms we can define a dissonant sound as one that is not pleasing to the ear. It might seem, therefore, that a dissonance is a type of tone we want to avoid. But this isn't true; both consonant and dissonant tones play an important role in music. The best way to see this is to analyze a musical score. Regardless of whether it is classical or pop, it is in many ways like a novel. In a novel the author attempts to create suspense, stress, and tension; then, near the end of the book he releases the stress and tension and—in many novels anyway—we have a happy ending. In the same way, a composer creates stress and tension using dissonance (or dissonant chords) in the center of a score and then resolves them using consonance.

Music of all types has become more dissonant in the last century. Stravinsky shocked many fans of classical music in 1913 when he introduced *The Rite of Spring*. With its ear-shattering dissonance and lack of conventional rhythm, it created a scandal. Many people were repulsed by it, others said they didn't understand it, but strangely, some welcomed it as a new revolution in music and admired Stravinsky's boldness. There's no doubt, however, that it shook the world of music down to its foundation. Dissonance is now also a standard part of jazz. As we will see in the next chapter, one of the mainstays of jazz is "superchords," which are generally very dissonant. It is, in fact, superchords that give jazz its distinctive tone.

Consonance and dissonance are obviously related to harmonics, or more exactly, the spectrum of harmonics, so let's look at how they are related. Consider the fifth; we know that it is one of the most harmonic combinations in music. Why? For an answer we have to look at the spectrum of harmonics of the two tones. Let's begin with the first tone (fig. 57). Its fundamental is f_1 and its overtones are $2f_1$, $3f_1$, . . . For the perfect fifth we have the spectrum shown in figure 58. Bringing them together, we get the spectrum shown in figure 59. We see immediately that several tones match up; furthermore, tones that do not match are a considerable distance from one another. Considerable separation of unmatched tones is important because tones close together will create beats, which are a problem.

We can also do the same type of analysis for the fourth, the third, and the other intervals. (On the piano the fifth is the interval C–G, the fourth is C–F, and the major third is C–E.) If we do this for all types of intervals and arrange the intervals in order of greatest to least consonance (i.e., dissonance), we get table 4, showing the greatest consonance at the top and increasing dissonance as you go down the column.

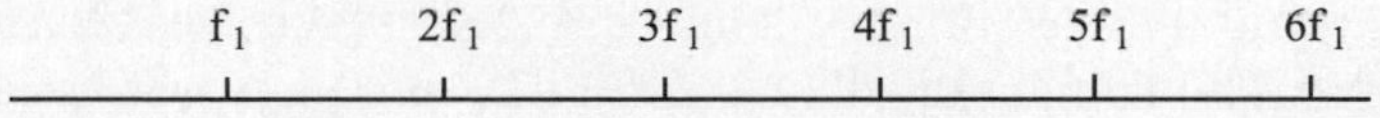

f_1 $2f_1$ $3f_1$ $4f_1$ $5f_1$ $6f_1$

Fig. 57. Spectrum of harmonics for the first tone.

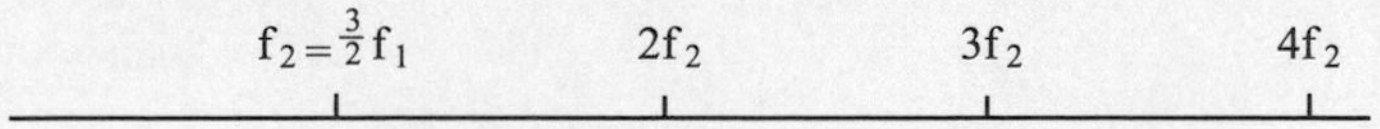

$f_2 = \frac{3}{2} f_1$ $2f_2$ $3f_2$ $4f_2$

Fig. 58. Spectrum for the perfect fifth.

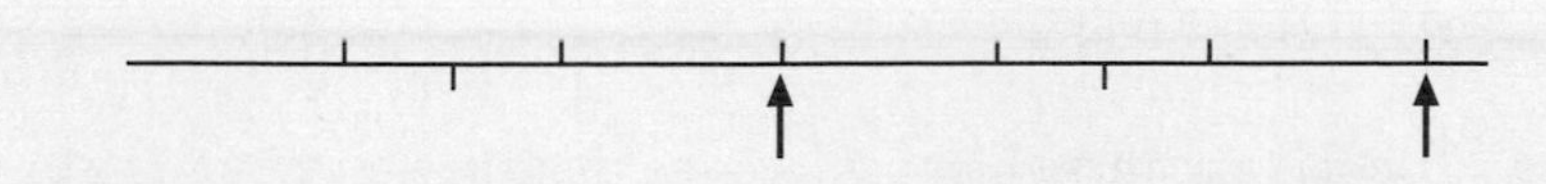

Fig. 59. Combined spectrum for the first tone and the perfect fifth.

Table 4. Intervals from greatest consonance to greatest dissonance

Interval	Notes (key of C major)	
Octave	C–C^1	Greatest consonance
Fifth	C–G	↓
Fourth	C–F	
Major third	C–E	
Major sixth	C–A	
Minor third	C–E♭	
Minor sixth	C–A♭	
Whole tone	C–D	
Half tone	C–D♭	Greatest dissonance

The Pythagorean Scale

Earlier, we saw that Pythagoras set up his scale using a hit-and-miss procedure with fifths. In this section I'll show you that it can be obtained in a more formal and logical way, but it will involve some mathematics. It is most convenient to use a step procedure, so I will use four steps. Like Pythagoras we will use the interval of a fifth; to obtain it we multiply or divide a fundamental, *f*, by 3/2. As we saw earlier, this sometimes gives a frequency that is in a higher or lower octave, and we must bring all tones into a single octave. We can do this by multiplying or dividing by 2 (or 4 . . . for higher or lower octaves).

Step 1. Starting from an arbitrary or frequency *f*, ascend in musical fifths (assume $f = 1$).

1	3/2	(3/2 × 3/2)	(3/2 × 3/2 × 32)	(3/2 × 3/2 × 3/2 × 3/2)	( )
1	3/2	$(3/2)^2$	$(3/2)^3$	$(3/2)^4$	$(3/2)^5$
1	3/2	9/4	27/8	81/16	243/32

This is the sequence of fifths. To obtain the actual frequencies associated with this sequence, you have to multiply the initial frequency by this number (e.g., in the C major scale, the initial frequency is that of C, which is 261.6 Hz).

Step 2. Looking at these numbers we see that they are not all in the same octave. If 1 represents the beginning of the octave, anything greater than 2 is in an octave above it. This means that in our sequence, 9/4 and everything above it is in an upper octave and must be brought into the main octave. In the case of 9/4 we can bring it down one octave by dividing by 2; this gives 9/8. Similarly, 27/8 can be divided by 2 to bring it into our octave; it becomes 27/16. The upper two numbers are in the octave above this, so they have to be brought down two octaves, and therefore must be divided by 4. When we do this, we end up with the sequence

1   3/2   9/8   27/8   81/64   243/128   2

But it is not in ascending order.

Step 3. Rearrange the notes so they are in ascending order. We get

1   9/8   81/64   3/2   27/16   243/128   2

If we now try to link these numbers to the notes C, D, E, F, G, A, B, and C—namely, the notes of the C major scale—we see that we are missing a note. We have only seven numbers, and there are eight notes in the scale (including the upper C).

Step 4. To obtain the missing note, descend from the starting note by a fifth, then bring the resulting note into the above octave. To descend we multiply by 1/(3/2) = 2/3; then to move up one octave, we multiply by 2. This gives 4/3. Our sequence is then

1   9/8   81/64   4/3   3/2   27/16   243/128   2

It is important now to look at the intervals between the notes, where interval means the ratio of the two notes. If we consider this to be the C scale, the interval (ratio) between C and D is (9/8)/1 = 9/8, the interval between D and E is (81/64)/(9/8) = 9/8, and so on. In this way we get the following sequence of numbers:

9/8   9/8   256/243   9/8   9/8   9/8   256/243

You can see that this sequence has only two intervals: 9/8 and 256/243. We refer to 9/8 as a *whole tone* and 256/243 as a *half tone*, so our scale is made up of a sequence of tones and half tones, namely, T T h T T T h, where *T* means tone and *h* means half tone. If we compare this sequence with the C major scale on the piano, we see that there is agreement: there is, indeed, no black note between E and F so the tone spread is smaller than between, say, D and E. A similar situation occurs between B and C.

Let's turn now to the accuracy of the scale. We know that the theoretical ratios for fifths, fourths, thirds, and sixths should be 3/2, 4/3, 5/4, and 6/5. Looking at the sequence we derived in step 4, we see that the fifth does have the value 3/2; similarly the fourth has the desired value 4/3. The third, however, is 81/64 and it should be 5/4, so it is slightly off. Also the sixth should be 6/5 and it is 27/16.

We can go further with the above approach. If we continue to lower and higher fifths we will get a scale that is made up entirely of half tones. We call it the *chromatic scale*. Again we will use a step procedure to obtain the scale.

Step 1. Using the previous numbers, continue ascending and descending in fifths.

$(2/3)^6$ $(2/3)^5$ $(2/3)^4$ $(2/3)^3$ $(2/3)^2$ $(2/3)$ 1 $(3/2)$ $(3/2)^2$ $(3/2)^3$ $(3/2)^4$ $(3/2)^5$ $(3/2)^6$

64/729 32/243 16/81 8/27 4/9 2/3 1 3/2 9/4 27/8 81/16 243/32 729/64

Step 2. Bring all the notes into the range of a single octave by multiplying or dividing by 2 and 4. This gives

1,024/729 256/243 128/81 32/27 16/9 4/3 1 3/2 9/8 27/16 81/64 243/128 729/512

Step 3. Arrange the notes in ascending order:

1,256/243 9/8 32/27 81/64 4/3 (1,024/729 729/512) 3/2 128/81 28/16 16/9 243/128 2

Looking at the intervals, we see that there are several different values, so the intervals are not all the same. Also, at the center there are two different values for the same note. So there are obviously serious difficulties with the scale.

The Just Diatonic Scale

Because of the difficulties we have just seen, a new scale was invented, the just diatonic scale, or more usually, the just scale (diatonic refers to a scale of seven notes in an octave). The just scale maximizes the number of consonant intervals; to do this, it starts with the major triad, which is made up of the tonic, the fifth, and the third. In the scale of C this is C-E-G, or in singing it is do-mi-sol. These notes are in the ratio 4:5:6. Again, we'll use a step procedure to set up the scale.

Step 1. Start with the triad f-$5/4f$-$3/2f$. With the common denominator 4 it's easy to see that these notes are in the ratio 4:5:6.

Step 2. Form a major triad up and down from the upper and lower numbers:

f $5/4f$ $3/2f$

$2/3f$ $5/6f$ f $3/2f$ $15/8f$ $9/4f$

On the upper triad we multiply by appropriate numbers so that the ratio of the three numbers is 4:5:6. The second number will therefore be $(5/4 \times 3/2)$ and the third $(3/2 \times 3/2)$. Similarly for the lower third, the three numbers also have to be in the ratio 4:5:6.

Step 3. Bring the notes from step 2 into the range of a single octave by multiplying or dividing by 2 or 4. For example,

> $2/3f$ needs to go up one octave, and becomes $4/3f$
> $5/6f$ needs to go up one octave, and becomes $10/6f$
> $9/4f$ needs to come down one octave, and becomes $9/8f$

Step 4. Arrange the notes in ascending order. We get

f $9/8f$ $5/4f$ $4/3f$ $3/2f$ $10/6f$ $15/8f$ $2f$

Finally, taking the ratios between neighboring notes gives

9/8 10/9 16/15 9/8 10/9 9/8 16/15

Again, there are problems with this scale. The first is that there are two different whole-tone intervals, namely 9/8 and 10/9. The half-tone interval is 16/15. Because we started with the major triad, the third and fifth are exact, but there are problems with the minor chords. We could set up a chromatic scale using this scale as we did in the case of the previous scale, but it has even more difficulties than the previous one. The major problem, however, is that the ratios down the scale are different, and therefore you could not transpose a piece of music from one scale to another properly. The tempered scale was set up as a solution to this problem.

The Tempered Scale

The ideal scale is one in which the steps between the notes are all the same and in which there are exact thirds, fourths, fifths, and so on.

But as it turns out, such a scale is impossible. So we have to compromise. We begin by making sure that all half-tone intervals are the same. In other words,

$$f(\text{note }1)/f(\text{note }2) = f(\text{note }3)/f(\text{note }2) = f(\text{note }4)/f(\text{note }3)\ldots$$

Also we require $f(\text{note }13) = 2f(\text{note }1)$. If we call the ratio of the half-tone intervals a, we have

Note 1 = $a^0 f$
Note 2 = $a^1 f$
Note 3 = $a^2 f$
. . . . .
Note 13 = $a^{12} f$

And since we have 12 equal intervals and 12 tones we can write

$$f(\text{note }1) = a^{12} f(\text{note one octave up}),$$

and therefore

$$a^{12} = \sqrt{2} = 1.0595.$$

This means that the tempered scale has the following frequencies, assuming we start with middle C = 261.6 Hz:

C	1.0000 × 261.6 = 261.6
C♯	1.0595 × 261.6 = 277.2
D	1.1225 × 261.6 = 293.7
D♯	1.1893 × 261.6 = 311.1
E	1.2601 × 261.6 = 329.6
F	1.3351 × 261.6 = 349.3
F♯	1.4148 × 261.6 = 370.0
G	1.4987 × 261.6 = 392.1
G♯	1.3878 × 261.1 = 415.4
A	1.6823 × 261.6 = 440.1
A♯	1.7842 × 261.6 = 466.3
B	1.8885 × 261.6 = 494.0
C	2.0008 × 261.6 = 523.4

This is, of course, a compromise with an exact scale, but in most cases the errors are not great and most people could not hear them. The percentage error in some of the intervals is given in table 5.

Table 5. Percentage error, selected intervals, in equal-tempered scale

Interval	Ratio		% Error
	Theoretical	Actual	
Fifth	1.5000	1.4987	.0877
Fourth	1.333	1.3351	.135
Major third	1.250	1.2601	.808
Major sixth	1.6667	1.6823	.936

The Major Scales

We now have enough information to discuss the major scales. Looking at the intervals in the tempered scale we see the sequence T T h T T T h, where, as we saw earlier, *T* is a tone and *h* is a half tone. Applying this to the scale of C major we get

Musicians have come up with the following names for these notes. They are also identified by Roman numerals. (Sometimes they are also referred to by arabic numerals.)

Tonic	Super-tonic	Mediant	Sub-mediant	Dominant	Sub-dominant	Leading note
C	D	E	F	G	A	B
I	II	III	IV	V	VI	VII

This sequence of tones and half tones is, of course, applicable to any scale. In other words, the tones can be applied to any notes. For the scale of G major, for example, if we apply T T h T T T h starting at G, we find that we get all the white notes except F, which is sharped. (Sharps are the black notes to the right of the note; flats are the black notes to their left.) The G major scale therefore has one sharp. If we apply this sequence to the F major scale we find we get one flat, namely B flat.

We can, of course, easily identify all the flats and sharps in a scale starting with C major, as shown in table 6. For the scales with flats, take the fourth note (in boldface) and bring it to the front. The

Table 6. Building the major scales from C major

Flatted scales							
C	D	E	**F**	G	A	B	C
F	G	A	**B♭**	C	D	E	F
B♭	C	D	**E♭**	F	G	A	B♭
E♭	F	G	**A♭**	B♭	C	D	E♭
A♭	B♭	C	**D♭**	E♭	F	G	A♭
D♭	E♭	F	**G♭**	A♭	B♭	C	D♭
G♭	A♭	B♭	C♭	D♭	E♭	F	G♭
Sharped scales							
C	D	E	F	**G**	A	B	C
G	A	B	C	**D**	E	F♯	G
D	E	F♯	G	**A**	B	C♯	D
A	B	C♯	D	**E**	F♯	G♯	A
E	F♯	G♯	A	**B**	C♯	D♯	E
B	C♯	D♯	E	**F♯**	G♯	A♯	B
F♯	G♯	A♯	B	C♯	D♯	E♯	F♯

fourth note in the new sequence will be flatted. You can continue with this for all the flatted scales, as the table shows. In this way we get the scales F, B♭, E♭, A♭, D♭, and G♭; and we can see the number of flats in each scale.

In the same way we can set up all the sharped scales, starting with C. This time we use the fifth tone and the seventh tone becomes sharped (we'll see the reason for this later). In this case we get the scales G, D, A, E, B, and F♯. If you play the notes of the last flat scale, G♭, and the last sharped scale, F♯, on a piano, you'll see that the notes you play are exactly the same, even though we can call them different names depending on whether we'll dealing with a flatted scale or a sharped one. This is called an enharmonic scale.

With this table we will always be able to identify the key that a piece of music is written in by looking at the number of flats or sharps.

The Minor Scales

Most music is written in major keys (scales), but they are not the only scales used. Another set of scales called the minor scales are also important. If you listen to a piece of music in a minor scale and compare it to one written in a major scale it's easy to hear the difference. Songs or compositions in major scales usually sound happy and upbeat, while those in minor scales are mournful, melancholy, or sad. Funeral marches, for example, are usually written in minor keys.

There are three minor scales: the natural minor, the harmonic minor, and the melodic minor; and each is distinguished by its tone sequence.

- The *natural minor* is made up of the following tones: T h T T h T T. If we apply this sequence of tones and half tones to the key of C we get the following notes: C D E♭ F G A♭ B♭ C. You can, of course, do this for every key.
- The *harmonic minor* scale is made up of the tones T h T T h (T+h) h. The only difference from the natural minor is that the sixth is raised by a half tone. For the scale of C this is C D E♭ F G A♭ B C.
- Finally, the *melodic minor* is made up of the tones T h T T T T h. The difference in this case is that the sixth note is raised by a half tone. For the scale of C we therefore have C D E♭ F G A B C. In the case of the melodic minor, descending on the same notes sounds awkward; because of this the descending notes are different. The tone pattern for the descending scale is T h T T h T T. In the case of the C scale this gives the notes C B♭ A♭ G F E♭ D C.

Relative Major and Minor Scales

You may have noticed in playing a piece with no sharps or flats that you thought was in the key of C major, that the music sounds as if it is in a minor key. Grieg's piano concerto is an example. It has the same notes as pieces written in C major, but it is actually written in A minor. This may seem strange, but you can easily hear the difference when you compare it with a piece that is written in C major. This applies to all scales; in fact, all major scales have a relative minor that has the same notes. Comparing A minor to C major we see that A is the sixth tone in the key of C, or alternatively you can think of it as three half tones down from C. If we look at other major scales such as F and G, we see that D is down three half tones from F and E is down the same amount from G. The D minor scale therefore has the same notes as F major, and E minor has the same notes as G major.

Pentatonic Scales

Pentatonic, or five-note, scales were introduced in the first chapter. They are used extensively in jazz, rock, and country music, and in the music of many countries around the world such as China and Japan. And any jazz pianist will tell you they are invaluable in improvising. Chick Corea and Herbie Hancock used them extensively.

Pentatonic scales can be obtained in several different ways, and as we will see, there are many different scales of this type. In chapter 1 we saw that Pythagoras obtained a pentatonic scale using fifths. Another method is to use the standard pentatonic tone sequence: T T (T+h) T (T+h). Looking at it we see that there are no individual half tones. In some of the variations of this scale, however, half tones are used. In fact, pentatonic scales are classified as either (1) *hemitonic*, containing one or more half tones; or (2) *anhemitonic*, containing no half tones.

Perhaps the easiest way to construct the major pentatonic scales is by eliminating the fourth and the seventh from a major scale. In the key of C this gives C D E G A; in the case of F it gives F G A C D, and for G it gives G A B D E. These examples might seem to suggest that most notes in pentatonic scales are white notes, but this is not true. In fact, the five black notes in any octave constitute a pentatonic scale. This brings us to another way of obtaining pentatonic scales: they are all the notes not in any given major scale. Since the C major scale contains no black notes, the sequence of black notes is a pentatonic scale.

Another interesting way of getting pentatonic scales comes from what is known as the "circle of fifths." As we will see later, this circle is invaluable in music. To form it we draw a circle and place C at the top. To the right of it we place the scale with one sharp (G), and in the next position at the right, the scale with two sharps (D), and so on. To the left of C we place the scale with one flat (F), and to the left of that the scale with two flats (B♭), and so on. The result is shown in figure 60.

As you move around this circle in a clockwise direction you see that each note is the fifth of its predecessor. If you move counterclockwise, the notes are separated by a fourth (e.g., F is the fourth of C). As it turns out, any five notes in sequence on this circle produce a pentatonic scale. For instance, we see C G D A E; rearranging these notes in the sequence of piano keys, we have C D E G A, which is the C major pentatonic scale. We also see that the five black notes constitute a pentatonic scale.

Many songs use only the notes of a particular pentatonic scale.

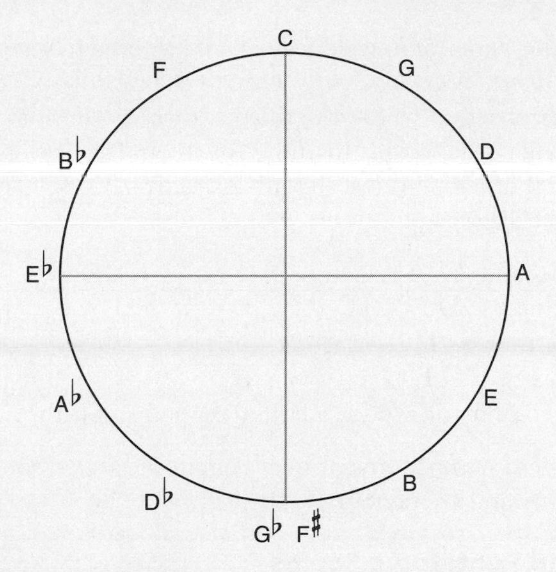

Fig. 60. The circle of fifths.

"Amazing Grace," "Auld Lang Syne," Chopin's "Black Key Etude," and Gershwin's "Summertime" are a few examples. You can easily check out "Amazing Grace" and "Auld Lang Syne" on the piano; in fact, it's relatively easy to transpose them so that they are entirely on the black note pentatonic scale. (I'll explain transposing later in the chapter.)

So far we've been talking about the major pentatonic scales. But there are several others; five of them are listed below, along with the major for C:

1. major (C D E G A)
2. thirdless with major sixth (C D F G A)
3. thirdless with minor seventh (C D F G B♭)
4. minor (C E♭ F G B♭)
5. fifthless (C E♭ F A♭ B♭)

As in the case of ordinary major scales, each of the pentatonic scales has a minor pentatonic with the same notes, referred to as the *relative minor pentatonic.* The tone sequence for the pentatonic minor is (T+h) T T (T+h) T. We know that C major pentatonic has the notes C D E G A; thus, A minor pentatonic has the notes A C D E G, which, rearranged, are the same.

The major pentatonic scales can also be altered in several ways to give us the *altered pentatonic scales.* One of the most important of this group is the variation where the third and the seventh are removed from the major scale rather than the fourth and the seventh (C D F G A). Several other altered scales are as follows (using C major pentatonic as our basis):

1. flatted second: C D♭ E G A
2. flatted third: C D E♭ G A
3. flatted fifth: C D E G♭ A
4. flatted sixth: C D E G A♭
5. various combinations such a flatted second and fifth: C D♭ E G♭ A

Although these are not conventional pentatonic scales, they still contain five notes and are pentatonic scales.

Modes and Pentatonic Scales

Modes are also used extensively in music. They are of Greek origin and therefore have Greek names. All the modes can be conveniently described by relating them to the major scales; for example, the Ionian mode is just the C scale. From here we step up one tone for each of the other modes.

C → C′ ($C_4 \rightarrow C_5$)	Ionian mode
D → D′	Dorian mode
E → E′	Phrygian mode
F → F′	Lydian mode
G → G′	Mixolydian mode
A → A′	Aeolian mode
B → B′	Locrian mode

These are the various modes for the scale of C, and it may seem that little new has been introduced. We have merely jumped up one note and played the same scale. If we look at these modes in more detail, however, we see there is a significant difference. Consider the Ionian mode (C → C′); it has the tone sequence T T h T T T h. But if we start on D and go from D to D′, we have the tone sequence T h T T T h T, which is quite different; it is the tone sequence for the Dorian mode. In the same way the tone sequences for the Phrygian and Lydian modes are h T T T h T T and T T T h T T h.

Let's apply the Dorian mode to C, in other words, apply T h T T

T h T starting on C. This gives the notes C E♭ F G A B♭ C, which is different from any scale we have seen so far. It is close to the minor scale, but is slightly different.

Now let's go back to the previous section where we defined five different pentatonic scales. We referred to them as major, thirdless major sixth, thirdless minor seventh, minor, and fifthless. If we look carefully we see that they are actually the

Ionian pentatonic
Mixolydian pentatonic
Dorian pentatonic
Aeolian pentatonic
Phrygian pentatonic

As an example, consider the thirdless with minor seventh; it was C D F G B♭. If we apply the Dorian mode to the pentatonic sequence T T (T+h) T (T+h) we get T (T+h) T (T+h) T, and applying this to C gives the notes C D F G B♭, which is the same as the above sequence.

The Blues Scale

The blues scale is particularly important in jazz and blues music, but it is also used in rock and gospel music. Ragtime—such as Scott Joplin's famous "Maple Leaf Rag" and the "Twelfth Street Rag"—is also based on the blues. Many of the early jazz greats were famous for their use of the blues. Among them were Jelly Roll Morton, Fats Waller, Louis Armstrong, and Art Tatum.

From a simple point of view, the blues scale is formed from the major scale by flatting the third, fifth, and seventh tones. In practice, however, the second and the sixth are usually left out. In the case of C this leaves us with

The flatted third and fifth notes are known as *blues notes.*

The blues scale can also be formed from pentatonic scales. It is the minor pentatonic with an added flatted fifth. Another way to form it is to start with a regular minor scale, omit the second and sixth tones, then add a note between the fourth and fifth. Finally, the blues scale has six notes and its tone sequence is (T+h) T h h (T+h) T.

Transposition

A song or piece of music is written in a particular scale, but you may have a singer who cannot reach the high notes in that scale. The way to get around this is to transpose the song to a lower key. For example, the song may be written in G; if you transpose it to C, all notes will be lowered by a fifth, since C is a fifth lower than G.

For example, suppose we have the following melody in G

Transposing it to C gives C → F, D → G, E → A and we have

Good musicians can transpose and play songs immediately, and it is a skill worth developing, but it takes considerable practice.

Other Scales

Several other scales are used in music. We have already met the *chromatic scale;* it is made up of half tones and is used extensively in jazz and other types of music. Its sequence is h h h h h h h h h h h h. Another scale that is sometimes used is the *whole tone scale*. As the name suggests, it is made up of whole tones (T T T T T T T). In the key of C we would have C D E F♯ G♯ A♯ C. Another scale is the *diminished scale*. It is made up of alternating tones and half tones: T h T h T h T. Finally, there are a number of *altered scales*. One example is a combination of the diminished scale and the whole tone scale with the sequence T h T h T T T.

6

Down Melody Lane with Chords and Chord Sequences

What would music be without harmony? Harmony, which comes from the playing or singing of several notes simultaneously, is one of the things that makes music beautiful. And for harmony we need chords. A chord is several notes—two, three, or more—played at once, and as we will see, chords play a major role in any type of music. Many musical instruments, such as trumpets and clarinets, play only a single note at a time, but others, such as the piano and the guitar, use chords extensively. Indeed, even though many instruments play only a single note, when they are in a band or orchestra, they are forming part of a chord. Instruments in an orchestra usually play different notes of a chord.

Dyads and Intervals

When two notes are played together they form a dyad, or interval. Several of these intervals are of particular importance because they are more harmonic than others (we have, of course, already met them). Because of their importance we refer to them as *perfect* consonances. They are shown in table 7. There are also several other dyads of importance—the *major* consonances—and they are shown in table 8. The smaller the ratio of numbers, the more harmonic the sound. In short, 3:2 is considerably more harmonic than 8:5.

Table 7. The perfect consonances

Perfect consonance	Frequency ratio	Examples in C major
Octave	2:1	C–C′
Fifth	3:2	C–G
Fourth	4:3	C–F

Table 8. The major consonances

Major consonance	Frequency ratio	Examples in C major
Major third	5:4	C–E
Minor third	6:5	C–E♭
Major sixth	5:3	C–A
Minor sixth	8:5	C–A♭

These intervals are illustrated in musical notation as follows

Chords

Triads

Three notes played together form a triad. We have already met the triad C-E-G and know that the frequency ratio of the three notes is 4:5:6; furthermore, from the previous section we know that the 4:5 ratio forms a major third and the 5:6 ratio forms a minor third. Our triad is therefore made up of a major and a minor third. In the same way, we can form several other important triads; they are listed in table 9. These are, of course, only a few of the triads that can be formed within an octave, but they are the most melodic ones.

Table 9. Some triads and their components

Triad	Components	Frequency ratio
C-F-A	Fourth + major third	3:4:5
E-G-B	Minor third + major third	5:6, 4:3
E-G-C′	Minor third + fourth	5:6, 3:4
C-E-A	Major third + fourth	4:5, 3:4
E-A-C′	Fourth + minor third	3:4, 5:6

When we talk about a triad, we refer to the bottom note as the root; in the case of the major triad, the upper two notes are the third and fifth. It's easy to see that the number of half tones (in the major triad C-E-G) between the root and the third is four. If we form a triad a tone up, such as

we see that there are only three half tones between the two lower notes, so it can't be a major. It is, in fact, a *minor chord;* in this case it is D minor.

If we continue up another tone we get the triad

and again, if we count the number of half tones between the lower two notes we get three, so this is also a minor chord. It is, in fact, E minor. The triad above it is F major, above that is G major, and above it is A minor. The last chord, which starts on B, is different. As we will see later, it is a "diminished" chord.

Earlier, we referred to the notes as we proceeded up the scale by the Roman numerals I, II, III, and so on. From the above we see that it is useful to also refer to the type of chord. If Roman numeral I is the tonic we write

I ii iii IV V vi vii°

where capital Roman numerals designate major chords, small numerals designate minor chords, and the degree symbol (°) designates a diminished chord.

Inversions

When a chord such as the C major triad has its root note on C, it is said to be in the root position. But you can position the notes differently and still have the same chord. For example, you could play them in the order E-G-C′ or G-C′-E. The alternate arrangements are referred to as inversions; the first is called the first inversion, and the second, the second inversion. We see that with three notes we have two inversions, and in general this is true. For any chord, the number of inversions will be one less than the number of notes in the chord.

Earlier we saw that in the root position we had a major and a minor interval. In the case of the first inversion (E-G-C′) we have a minor third and a fourth, and for the second inversion (G-C′-E) we have a fourth and a major third.

Tetrads

If the chord consists of four notes rather than three, it is referred to as a tetrad. The most common tetrad is the *seventh.* There are two basic types of sevenths, the major seventh and the dominant seventh. In practice, the dominant seventh is used more frequently than the major seventh and is usually referred to simply as "the seventh." The note added in the case of the major seventh is a half tone below the upper key note; in the case of C, this is B. For the dominant seventh the added note is a tone below the upper key note; in the case of C this is B♭.

The major seventh produces a slightly dissonant tone that jazz musicians like, so it is frequently used in jazz.

Along with the seventh we have the *sixth* chord. It is also used extensively and is formed by adding a key one tone up from the fifth to the triad; in the case of C, this is A. Sixth chords play an important role in popular music, and many jazz pianists add them to all (or at least, most) major chords.

Augmented and Diminished Chords

In addition to the types of chords mentioned above, two others are of importance. We have, in fact, already met one of them. They are the *augmented* and *diminished* chords. The augmented is formed by sharping the fifth; the diminished is formed by flatting the third and fifth. Both types of chords are used extensively but are not as common as major, minor, or seventh chords. In some cases a seventh is added, so that you have an augmented seventh or diminished seventh. Although both sevenths are used, the dominant seventh is more common.

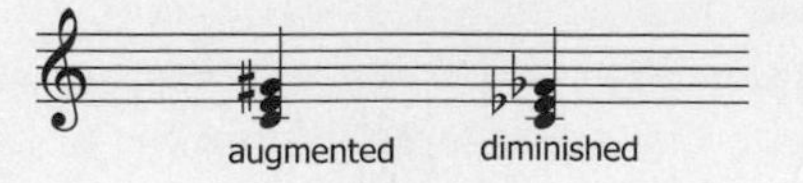

Table 10. Chord notation used in sheet music

Chord type	Symbol
Major	C, Cmag, CM
Minor	Cmin, Cm
Augmented	C^+, Caug
Diminished	Cdim, C^o
Seventh (dominant)	C7, C^7
Major seventh	Cmaj7
Minor seventh	Cmin7
Diminished seventh	Cdim7, C^o7
Augmented seventh	C^+7, Caug7

Notation

Since we now have discussed several types of chords, let's turn to how we represent them in sheet music. Popular sheet music usually provides the chords associated with the melody. For the most part the notation is standard, but there are deviations. Table 10 shows the usual chord abbreviations, along with a few of the deviations.

Extended and Suspended Chords

Before we look at extended and suspended chords, I would like to briefly mention another addition to chords. In any chord, it is common to add the upper octave note. For example, if you were playing the triad C-E-G, it is common to add in C′, so that the chord becomes C-E-G-C′. For people with small hands, this chord may be a problem, but as we will see later, it is very helpful in playing melodies.

In extended chords not only is the upper octave note added in, but notes beyond it can also be added. Chords of this type are sometimes referred to as *superchords.* The most common superchords are the ninth, the eleventh, and the thirteenth. Again, C can be used to illustrate the chords. In the case of the ninth we add in the D above C′; for the eleventh we add the upper F; and for the thirteenth we add the upper A.

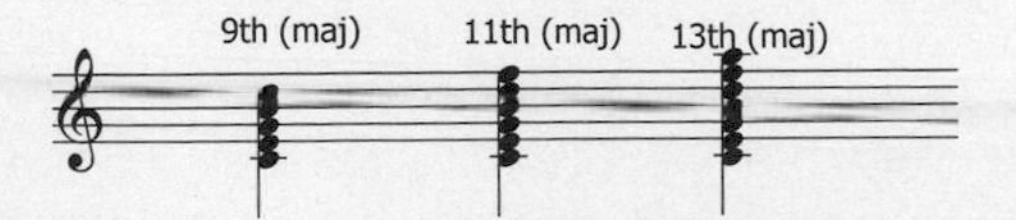

It might seem impossible to play these chords; after all, we only have five fingers. In practice, however, the extra notes are usually played in the lower octave, or some of the notes of the major chord are left out. In the case of the ninth, for example, the D above the lower C would be played, and in the case of the eleventh, F would be played.

The notation for these chords is C9, C11, and C13. You've probably noticed that the upper notes can be played with either a major seventh or a dominant seventh. The above notation usually applies to the dominant seventh; for the major seventh we use Cmaj9, Cmaj11, and Cmaj13.

Closely associated with extended chords are suspended chords. There are two types: the suspended second and the suspended fourth. The suspended fourth is the more common of the two. In the suspended fourth the third is raised by a half tone; therefore, for the C triad, instead of C-E-G, we would have C-F-G. If you play this chord on the piano it may sound slightly dissonant to you, but it is a popular chord in jazz. For the suspended second, you lower the third by a tone; in the case of C this gives C-D-G. You will no doubt ask how these chords are different from the ninth and eleventh. They do have the same lower notes, but because the suspended chords do not have sevenths and so on, they are distinct. The symbols for suspended chords are Csus4, Csus2.

Filling in the Melody

The question now is, What do we do with all these different chords? The major thing we use them for is filling in the melody. When you buy popular sheet music, you get a melody line for the song along with the chord symbols for each measure. There is also usually a simple arrangement of the song. If you read music, you've probably played the printed arrangement, but for most musicians it is usually too simple. What I'd like to do is show you how you can fill in the melody with chords and make up your own arrangement of the song. Let's consider a relatively simple song such as "In the Good Old Summertime." It is written in the key of C and is in 3/4 time.

The sequence of simple notes gives us the melody, and the symbols above give the chords. As you might expect, there are many ways of filling in the chords. I will outline the traditional way in a series of steps, then briefly discuss some of the variations.

1. Raise the melody an octave higher than it is written on the sheet music. This is to avoid conflict with the left-hand chords. (If the melody is already relatively high, you may not want to do this.)
2. Play the first few notes (the lead-in notes), such as those in the musical example above, singly or as octaves.
3. Play the first note in each measure as the top note of a chord. In other words, the chord is played *down* from the melody note. Use the appropriate chord according to its symbol above the bar (e.g., C-F-G).
4. If the song is relatively simple, and contains few eighth notes, you can play most of the notes as chords. On the other hand, if the piece is fast, and contains many eighth notes, it's usually impossible to chord them all; even if you did, the music would sound too dense.
5. If you are not adding a chord to a note, you play it as a single note or as an octave.
6. If the melody note is not part of the usual triad (or tetrad), play it along with as many notes of the triad as you can.

Applying these steps to the above song gives the following rendition:

In practice you won't be reading notes in the upper register because everything is done at the piano, and you are merely moving up one octave. You can, in fact, get around thinking that the melody is an octave up by playing all chords with their extended octave note. C major chord, for example, would then be C-E-G-C′. You can then think of the chords as extending *upward* from the melody note. Thus, the first four bars would be as follows:

This will give you a few more notes and make the arrangement sound fuller. Actually, in most cases, you'll find that all these complete chords make it sound too full. To get around this it is best to play some of the notes as octaves, or even as single notes.

What About the Left Hand?

Many different patterns are used in the left hand when it is accompanying the right-hand melody. The simplest is a sequence of single notes. The first note in the bar should always be the root note of the chord; other notes can be the third or fifth of the chord. In the above song you might have

Or you could use part or all of the of the cho

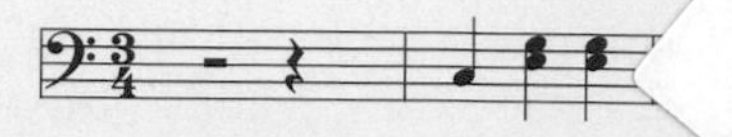

You can also use more complete chords; in this
priate to play the first note lower.

If the song is in 4/4 (or common) time rather than 3/4 time, the bass pattern will, of course, be different. One of the most common patterns is referred to as the *swing bass.* In this case you begin with the root note, play a chord, then play the fifth above the root note, and finally the same chord again, as below.

For variety, you can, of course, change this in several ways. The single notes can be octaves, or you could play sequences of chords as below, or change the order of the chords within the bar as shown.

Needless to say, there are many possibilities.

Voicing and Arranging

Playing everything with blocked chords and a few single notes along with a swing bass would soon get boring, and everything you played would sound the same. To spice things up you have to use variety. In particular, you don't have to use standard triads for every chord; the idea is to select the best-sounding chords. This selection is usually referred to as *voicing*. In some cases, where there are normally four or five notes in a chord, you may want to use only two or three. You may choose to use a sequence of dyads or single notes for a bar or two. The trick is to experiment and find out what sounds the best. This also applies to the left hand; there are numerous left-hand variations that you may want to use. I discussed a few of them earlier, and you will likely want to experiment with others. An example of this for the song "In the Good Old Summertime" is as follows:

In all songs there are "dead" spaces, or measures where there is little going on (long melody notes). This gives you another opportunity to be creative. There are numerous types of "fill" that can be added. Examples are a simple counter-melody of a few single notes, a sequence of intervals or dyads, an arpeggio of chord notes, or simply some chords. These fills are sometimes played in the octave above the melody.

The object is to find an arrangement that you like and memorize it. You can always vary it slightly when you're playing it, but you now have a foundation to work with.

Chord Sequences and the Circle of Fifths

The circle of fifths was introduced earlier, and I mentioned that it is invaluable to musicians, particularly those dealing with chords. To help you remember I show it again in figure 61. As I stated earlier, it's a good idea to memorize it.

The circle of fifths is particularly important in relation to the sequence of chords in a song. As we saw in "The Good Old Summertime," the chord changes every few bars, or sometimes within a

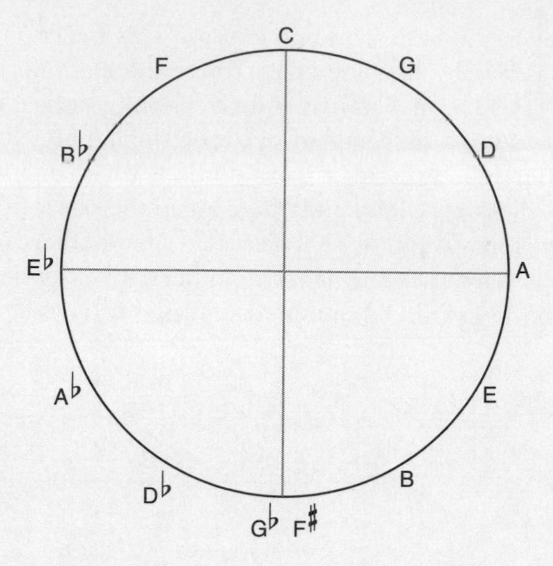

Fig. 61. The circle of fifths.

single bar. Over the first few bars it had the chord sequence C C7 F C, and as you look at other songs, you will see similar sequences. So the question is, Do these chords change in a regular way? In other words, is one particular sequence preferred over another? As it turns out, chords do, for the most part, change in a particular way. If you look at the sequence above, you see that changes are gradual; in other words, most of the notes in a given chord of the sequence are the same as in the chord prior to it. This tells us that if chords change too drastically, things don't sound right. Rarely, for example, would you see a sequence like C A♭ B C. The reason is that successive chords in this sequence share few notes.

The circle of fifths is useful because it allows us to determine acceptable sequences, and it helps us understand where they come from. Most standards and older folk songs have relatively simple sequences. If the song is in C, for example, it might, over the first eight bars, have the sequence C F G7 C. Let's use the circle of fifths to analyze this (fig. 62). We'll draw an arrow from the center up to C, which represents the key of the song, or "home base," as we'll call it. We see immediately that in going to F and G, the chords are not varying much from the key chord of C. In fact, our arrow is always drawn back to home base.

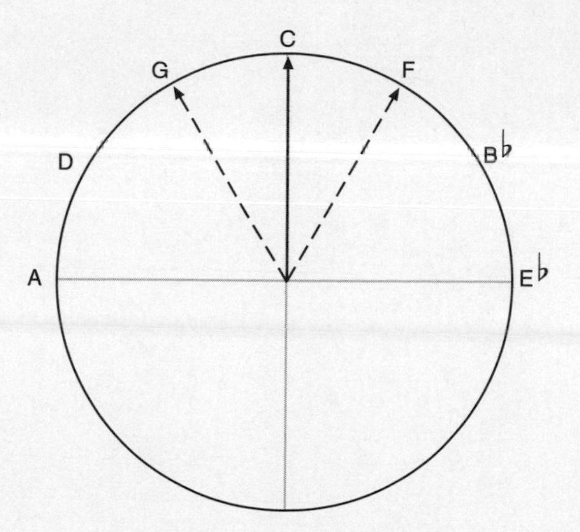

Fig. 62. Circle of fifths showing a song with a sequence C F G^7 C.

In more modern songs the song always starts out on the key chord, but then goes to a chord that is two or three, or perhaps more, places to the left (counterclockwise) of the chord. In the key of C, this might be A. But it is always attracted back to home base, so it will then move to a D chord, then to a G chord, and finally C (fig. 63). The question now is, What types of chords are used? For an answer to this we have to go back to our Roman numeral representation of the chords:

C	D	E	F	G	A	B
I	ii	iii	IV	V	vi	vii°

This tells us that in C major whenever we have a D, E, or A chord, it should be a minor; IV and V chords (F and G), on the other hand, should be major chords. But if we use seventh chords for the major, we'll probably share more notes with the previous chord. Our sequence is therefore C Dmin G7 C. We can also represent this sequence as I ii V I; most of the time, however, the first symbol (I) is usually left out because all sequences begin on I.

Let's consider a song in E♭ to illustrate chord sequences further. In this case the home base is E♭, and again our arrow moves several chords counterclockwise (fig. 64). Assume that it moves a quarter of a circle to C; it will then start making its way back to E♭, going first

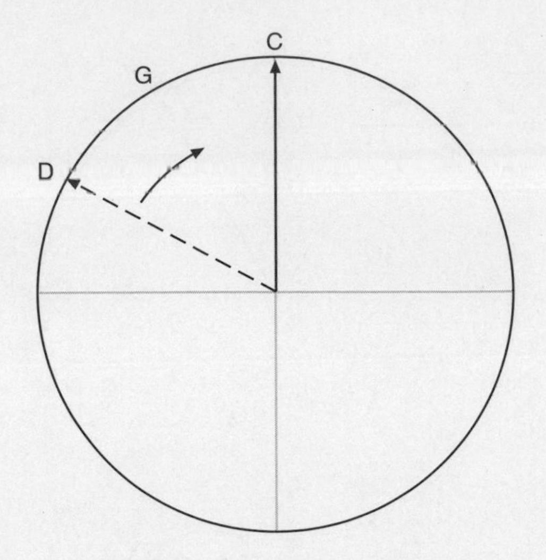

Fig. 63. Circle of fifths showing attraction back to the key note.

through F, then B♭. Now for the E♭ scale we have the following lineup between the notes and the Roman numeral designations:

E♭	F	G	A♭	B♭	C	D
I	ii	iii	IV	V	vi	vii°

Using this we can determine what chords we need. F is a minor chord, so we use Fmi, C is also a minor chord, and B♭ is a major chord so we will use B♭7. Our sequence is therefore E♭ Cmi Fmi B♭7 E♭. In some cases the minor chords will also have added sevenths.

Finally, let's look briefly at a sequence to the left of C, say one in the key of G. The home base is G, so the arrow starts on G, as shown in figure 65; it then moves several places counterclockwise. We'll assume it moves to E; from here it starts its homeward motion, first to A, then to D, and finally to G. A and E will both be minor chords, and D will be a seventh.

For most songs this is the way the chords progress, but there are no rigid rules, and many songs deviate from it. They may move two places counterclockwise from the keynote, or they may move four, but they almost never move more than a half circle. Also, on their route home they may decide to move back for a bar or so, but eventually they head home again.

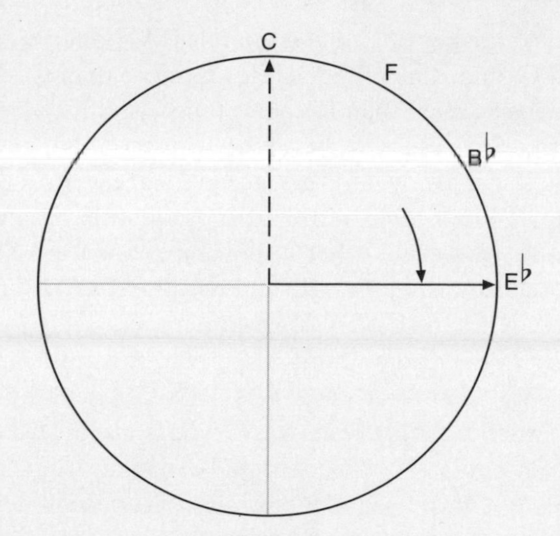

Fig. 64. Circle of fifths for a song in E♭.

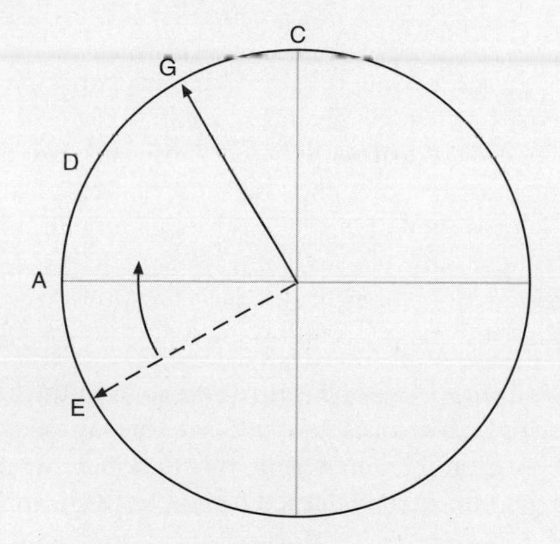

Fig. 65. Circle of fifths for a song in G.

We can obtain chord sequences using the circle of fifths as above, but if you've read any books on jazz you are likely to have seen the II V I sequence referred to again and again. In this case Roman numerals are being used to designate the sequence, and this is quite common. For the key of C, the numerals II V I refer to the sequence C Dmi G7 C, but in reality they refer to more than this sequence. If we use numbers rather than Roman numerals, this sequence is 2 5 1, but using the circle of fifths we can easily extend this to 4 7 3 6 2 5 1. In this case the arrow would start at 180 degrees from the key note. The numerals II V I refer to this entire sequence, or any part of it.

There are, of course, other important sequences. The sequence IV V I is used extensively in jazz and the blues, and other sequences such as V IV V I and III VI V I are sometimes used.

You may wonder how all this is useful to the musician. As it turns out, it is useful in several ways. First of all, it gives him some idea what the next chord is likely to be when he is playing a song. It is also useful to composers; they obviously have to know what chords to use. A knowledge of chord sequences is also particularly helpful in harmonizing a song, and it is useful if improvising (free playing), which as we will see, is an important part of jazz.

Chord Substitutions

Most songs are written using relatively simple chords: mostly major and minor chords with a few sevenths and an occasional diminished chord. Many musicians, particularly jazz musicians, like to "jazz up" the piece using more complex chords, so they make substitutions for many of the chords shown on the sheet music. Again, there are no rigid rules for these substitutions, and the player has to use her ingenuity, but there are a number of standard substitutions many musicians use. Some of them are as follows:

- *Tritone substitution.* A tritone is three tones from the key chord. In the case of G, this takes you to C♯ and means that you can substitute C♯7 for G7. The easiest way to remember this is to note that the tritone is the flatted fifth. This is a favorite of jazz musicians.
- *Chord extensions.* You can always add notes to any chord. A common substitution is to play C6 for C. A major seventh, a ninth, or an eleventh chord can also be used. Suspended chords such as Csus2 or Csus4 can also be used as a replacement for C.
- *Relative majors and minors.* We saw earlier that for every major scale there is a relative minor scale, a third below the major,

with the same notes. The key of C, for example, has Amin as its relative minor. The relative minor chord can be substituted for its major. Interestingly, the minor chord a third above the major is also a good substitution. In the case of C this is Emin.

- *Diminished and augmented chord substitutions.* Any chord that shares several notes with the original one can be substituted. You can therefore experiment with substituting diminished and augmented chords for major ones. The only important thing is that it sounds right. A common substitution is a diminished seventh chord a third above the chord you are using.
- *Steps substitutions.* The chord of a given note can be "stepped into" using the chord next to it. For example, C G7 can become C F♯ G7.

There are, of course, many different substitutions—too many for me to describe here—and the best way to find them is through experimentation.

Improvising

Earlier, I mentioned that improvising is an important part of jazz and a useful skill to develop. In many ways it is an art, and there are no rules or recipes for becoming an accomplished improviser, but there are techniques that help. For the most part, improvising is based on scales and chord arpeggios (the notes of a chord played one after the other). The chords themselves are also used extensively. Pentatonic scales are particularly useful, as are blues scales. "Accidentals" are frequently used along with the scale for variety.

The best improvisers use what are called *motifs,* four-, five-, or six-note sequences, similar to those that occur in most songs. If you watch these performers carefully, you see that they link motifs together in various ways. One of the best strategies, in fact, is to memorize five or six motifs, and then play them in various ways. Play them forward, play them backward, leave out notes, add a few extra notes here and there. These motifs are frequently based on scales and chord arpeggios, and indeed, you can link them with sections of scales and chord arpeggios. The addition of a series of chromatic notes is also very effective.

The overall form of most songs is AABA or ABAB (where A and B represent repeating sections), with A and B being 8 bars long, for a total of 32 bars. Sections in the blues, on the other hand, are usually 12 bars long.

A knowledge of chord sequences is essential in improvising. You can go back as far as 36,251, or just use 251 (in C these would be the sequences Emin Amin Dmin G7 C, or Dmin G7 C). It's important, however, to be able to improvise in any key.

7

"You've Gotta Have Rhythm"

Rhythm and Types of Music

Rhythm plays an important role in music. One person who had a lot was George Gershwin. "I've Got Rhythm" was one of his most popular songs. Belted out by Ethel Merman in the Broadway play *Girl Crazy*, it was the hit of the show and went on to become a popular standard. Rhythm is, indeed, one of the three major ingredients of any piece of music, along with melody and harmony. Rhythm is particularly important because, to a large degree it determines the type of music. As anyone who listens to music knows, rock music has a very different rhythm from New Age or classical.

What exactly is rhythm? Formally, it's defined in the *Concise Oxford English Dictionary* as "the feature of musical composition that is concerned with the duration of notes and periodic accent." For most people, however, rhythm is associated with the beat you tap your feet to or clap your hands to. It is related to the underlying pulse of the music and involves patterns of time. A rhythmic unit is a pattern of notes that occupies a certain period of time and is repeated over and over. In modern music, rhythm is usually emphasized by a rhythmic section consisting of drums and a bass, or sometimes instruments such as keyboards and guitars. They emphasize the beat and keep it going. The speed of the beat is referred to as the *tempo*.

Syncopation and anticipation frequently play an important part in rhythm. Syncopation is accenting parts of the beat that are usually not accented; anticipation is playing the note before the beat. So, for instance, the last eighth note in the first measure here anticipates (and replaces) playing the note as beat 1 of the second measure:

One form of syncopation is putting the emphasis on the *backbeat;* it is particularly important in rock and roll and reggae music. In 4/4 music the backbeats are the second and fourth; the first and third beats are referred to as the downbeats, which are usually emphasized in classical music. Backbeat syncopation is usually supplied by drums or bass.

As I mentioned above, it is usually the rhythm that determines the type of music. In this chapter we will look at the various types of music and the rhythms associated with them. There are, of course, many different types of music, and I won't be able to cover them all, so I apologize if your favorite type is not mentioned. In the next section I'll give an overview of the major types of music (in no particular order, I should mention). In the subsequent sections we will look at each in detail.

Overview of Types of Music

I'll start our overview of types with rock and roll music (frequently called rock 'n' roll). It burst onto the scene in the late 1940s and early 1950s and has been popular ever since, particularly among younger people. Its dominant feature is a strong beat, which, in turn, makes it "danceable" music, and this is, of course, one of the reasons it has always been popular with the younger crowd.

The second type of music on my list is the blues. It traces back to the religious spirituals of early African American slaves. As the name suggests, the lyrics are usually concerned with problems, difficulties, and troubles. It is based on the blues scale and "blue notes," with flatted thirds, fifths, and sevenths, which give it a sorrowful or blue sound. As we will see, it has influenced many different types of music, and while it has changed over the years, it still has the same basic features.

Boogie-woogie is closely associated with the blues. While it can't be considered a major type of music, it has influenced many genres. It is generally upbeat and fast, with the left hand carrying the rhythm. Boogie-woogie had its heyday in the 1930s and 1940s, but it is still heard occasionally today, even on some of the hits.

One of the most truly American styles of music is jazz. It's a little difficult to define, mainly because it has so many subgenres. Ragtime, Dixieland, bebop, and several others are all special types of jazz. Jazz is characterized by blue notes, so it's obviously associated with the blues. It is very rhythmic and makes considerable use of swing, syncopation, anticipation, and polyrhythms (more than one rhythm at a time). An important aspect of jazz is free-playing, or improvisation. All the early jazz musicians were great improvisers. Superchords such as ninths and elevenths also play an important part in jazz.

Country music may seem a long ways from jazz and the blues, but it does have ties with them, and it certainly has ties with rock and roll. There is, in fact, a branch of country that is referred to as country rock. Rhythm is important in country music, but it is not emphasized as much as it is in rock and roll. The electric guitar is now the major instrument of this genre, but other instruments are, of course, used. The music is generally upbeat, but like the blues, the lyrics are frequently about problems related to love.

The audience for New Age music is generally much smaller than that of country or rock and roll. Many who enjoy it are also strong believers in the New Age lifestyle. Some, however, are not. Beat and rhythm are not important here. New Age is generally smooth music; many people, in fact, refer to it as "nature" music; and indeed some of the sounds of nature such as flowing water, wind, waves, and waterfalls are occasionally incorporated into New Age recordings.

We now come to a catchall label—"pop music." Commercially, pop is no doubt the biggest seller and is generally directed at teenagers, as they buy large numbers of CDs. Pop music often includes music from the other categories discussed here, particularly rock and roll (or just "rock" as it is usually called now). Pop music has changed significantly over the years; the pop music of the 1940s and '50s was quite different from that of the 1980s, '90s, and beyond. Over the years faster music with a heavier beat gradually increased in popularity.

Rhythm and blues is an offshoot of blues and jazz. It goes back to the early 1940s and was a predecessor to rock and roll. In its early form it was strongly influenced by "jump blues" and early African

American gospel music. Today it is usually associated with "soul" music and funk (a fast, rhythmic type of music that deemphasizes melody and harmony).

Gospel music came from the Deep South and early on was associated with African American churches. It is usually divided into slow gospel and fast gospel.

Reggae music, from Jamaica, is characterized by a strong backbeat. Although it is a type of Latin American music, it is quite different from most of the other styles from the region, such as the rumba and tango.

Finally, at the end of the chapter we look at classical music, not because I feel it is the least important. It certainly isn't. It is the earliest form of music, and all the other types derive from it to some extent. There are several divisions within classical music that are associated with eras, and I'll talk about each in detail and also about some of the better-known compositions.

Rock 'n' Roll

Rock 'n' roll began in the late 1940s. It was derived from jazz, the blues, and even early country music; an early form, in fact, was called "rockabilly," which can be traced to what was called "hillbilly" music. The origin of the term *rock and roll* is uncertain, but some of the early gospel songs of the South had titles with the word "rocking" in them. The Cleveland, Ohio, disc jockey Alan Freed is sometimes credited with the name, and there's no doubt that he put rock and roll on the map. He organized the first of several rock concerts in 1952. Huge crowds attended them, and they were generally considered to be a tremendous success. The early ones were attended mainly by African Americans, but soon more and more whites started to come, and rock and roll really took off.

One of the earliest hits that can be considered a rock and roll hit was Joe Turner's piano single of 1939, "Roll 'em Pete." In 1954 he hit the charts again with "Shake, Rattle, and Roll." And of course there was Elvis Presley, who burst onto the scene in 1954 with his hit "That's All Right (Mama)." And as everyone knows, it was followed by a string of other Presley hits. Also about this time came the incredibly popular "Rock around the Clock" by Bill Haley and the Comets. It topped the charts in the United States for weeks on end and also became a hit in England and Australia.

Other important rock and roll artists of the time were Chuck Berry and Little Richard. Little Richard exploded on the scene with his "Tutti Frutti" in 1955. Over the next few years he recorded 50 songs, releasing two albums and nine singles. He combined boogie-woogie piano with a heavy backbeat and gospel-like lyrics. Many of the rock musicians who followed him claimed to be strongly influenced by his style. One of them was Jerry Lee Lewis, who like Little Richard, was primarily a pianist, and Buddy Holly, whom many refer to as the King of Rock and Roll. Holly was killed in a plane crash in 1959.

All of this set the stage for the "British invasion," and among the first to come to America were the Beatles: John Lennon, Paul McCartney, George Harrison, and Ringo Starr. After an appearance on the *Ed Sullivan Show* they gained international fame, and over the next few years they released over 40 singles and albums that reached number one on the charts, making them one of the best-selling bands of all time. It is estimated that they sold well over a billion disks and tapes.

One of the most distinguishing features of rock and roll is its beat. It has an accented backbeat, usually provided by snare drums. The piano was featured in many of the early rock and roll hits by Little Richard and Jerry Lee Lewis, but it was soon replaced as the lead instrument by the electric guitar. Saxophones were also used in many of the early hits but have not been used much in later rock and roll. Keyboards, however, are still used extensively.

So, what characterizes rock and roll? Early rock and roll frequently had an eight-beats-to-the-bar structure similar to that of boogie-woogie. A typical pattern is

Other patterns that are commonly used in the bass are solid and broken octaves as follows:

Single repeated notes are also frequently used.

Anticipation is important, as illustrated in the following pattern

Anticipation is also frequently used in the right hand.

Simple bass patterns sometimes become very repetitive; a good variation on this that is sometimes used is

One of Jerry Lee Lewis's favorite left-hand patterns was

While these patterns look similar to the boogie-woogie patterns, they are played differently in rock and roll. In boogie the emphasis is usually on every second note; in rock all notes are emphasized equally.

The Blues

Rock and roll owes a lot to the blues. In fact, the blues have had a strong influence on many types of music, including jazz, Dixieland, bluegrass, rhythm and blues, country, and pop. The earliest blues songs were based on the "call and response" that was used by slaves as they worked in the fields; a similar device was used earlier in Africa. In call and response, one group sings a phrase, then another responds. The lyrics are usually about personal woes, hard times, cruelty, lost love, oppression, and misery.

The blues first appeared in about 1900, and by 1912 Tin Pan Alley had published several blues songs. One of the earliest was "Memphis Blues" by the talented black musician W. C. Handy. Handy later went on to write the popular standard "St. Louis Blues."

In the 1940s the "jump blues" became popular. This style introduced a jazzy, up-tempo sound and employed many instruments that had not been used in the blues earlier, such as the saxophone and guitar. Other brass instruments such as the trumpet were also starting to be used. Louis Jordan and Big Joe Turner published jump blues tunes that later had an influence on rock and roll.

Louis Armstrong, Duke Ellington, B. B. King, Muddy Waters, Miles Davis, and even Bob Dylan all made important contributions to both the blues and the jump blues. And of course, both Elvis Presley and Bill Haley fell under their influence. Although some of Elvis's early songs were referred to as "rockabilly," many musicians said they were actually "blues with a country beat." Indeed, rock and roll has been occasionally referred to as "the blues with a backbeat."

Some of the earliest rock and roll songs that were strongly influenced by the blues were "Johnny B. Goode," "Blue Suede Shoes," "Whole Lotta Shakin' Goin' On," and "Shake, Rattle, and Roll." Surprisingly, the blues even had an influence on what is now considered classical music. George Gershwin's "Rhapsody in Blue," introduced by band leader Paul Whiteman in 1924, combined elements of classical music with blues and jazz and was an immediate success. Indeed, it went on to become a popular standard that is still played at many concerts today.

Twelve-bar blues became the standard in the early 1930s. It typically consisted of three sets of four bars with the sequence I IV I V IV I. In the key of C this would be C (4 bars), F (2 bars), C (2 bars), G (1 bar), F (1 bar), C (2 bars). Usually the last two bars were left for improvising, which was a major part of the blues.

The blues relies heavily on blue notes—the flatted third, fifth, and seventh—and "crushed" graced notes are used extensively, along with triplets. Many different rhythms are present in the blues; boogie-woogie basses such as that below are quite common.

Boogie-Woogie

Boogie-woogie basses are so important in both the blues and jazz, and even in rock and roll, they are worth looking at in some detail. These bass configurations are usually associated with the piano but can be played on other instruments such as the guitar.

Boogie-woogie became popular in the late 1930s and early '40s. It seems to have originated in the logging and turpentine camps of the South, and to some degree in the oil boomtowns in the early 1900s. Early on it had many forms, and it wasn't referred to as boogie-woogie until the late 1920s. It was, in fact, sometimes referred to as "eight to the bar" because it was usually played in 4/4 time with eight eighth notes to the bar.

One of the earliest boogie-woogie hits was "Pinetop's Boogie-woogie" by Pinetop Smith. Clarence Williams of New Orleans is credited with making the first boogie-woogie phonograph record in 1923. It was, in fact, during the 1920s that the style began to spread and become popular. George Thomas's recording of "The Fives" in 1923 introduced right-hand patterns that soon became boogie-woogie staples. Even today, musicians who play boogie-woogie frequently use the right-hand patterns from "The Fives." The left hand used one of the more popular patterns, called "walking octaves," which are broken octaves. Later boogie-woogie hits included "Honkie Tonk Train Blues," "Swanee River Blues," and Joe Turner's "Roll 'em Pete." "Roll 'em Pete" was played in Carnegie Hall in 1937 when Big Joe Turner played a tribute to Pete Johnson.

Some of the more popular bass patterns of boogie-woogie are the following:

Jazz

Most people who write about jazz say that it is difficult to define, and indeed it is. It has many different characteristics and forms. Like the blues, it uses blue notes extensively; it also uses syncopation, anticipation, swing, triplets, call and response, and polyrhythm, but its most important characteristic is its strong beat, or rhythm.

Jazz has its roots in the folk music of the slaves of the South. It was also heard frequently in early New Orleans, particularly during funeral processions. An early form of jazz was ragtime. About 1900 Scott Joplin, the son of a former slave, who was schooled in classical music, wrote some of the first ragtime hits. Two of his most popular—"The Maple Leaf Rag" and "The Entertainer"—are still popular today and frequently played at concerts. Ragtime even spilled over to Tin Pan Alley, with Irving Berlin publishing his "Alexander's Ragtime Band" in 1911.

Another popular branch of jazz was Dixieland. It developed in New Orleans about 1900 and soon spread to New York and Chicago.

It is sometimes referred to as the first "true" jazz and was the first music referred to as jazz (in about 1913). Various instruments were used in Dixieland: trumpets, clarinets, guitars, piano, bass, and drums. It was usually played by small bands, and much of the music that came out of them was improvised. Louis Armstrong's band is usually identified with Dixieland.

Several larger orchestras and bands, in fact, became known as jazz bands. Duke Ellington's band was one of the first, and Ellington himself wrote many of the early jazz standards. Interestingly, Ellington required his musicians to be formally trained in music, even though much of jazz was improvised and played by musicians who had little formal musical training. Another great early jazz band was that of Paul Whiteman. As we saw earlier, he introduced Gershwin's "Rhapsody in Blue"; he also introduced many other jazz standards. He eventually became known as "The King of Jazz."

The invention of the phonograph record revolutionized the music industry, and there's no doubt that it helped jazz. Records soon became very popular, and during the 1920s radio stations began playing jazz hits along with other popular music. The 1920s, in fact, became known as the "Jazz Age."

In the 1930s jazz started to give way to "swing." In many ways, swing was just another form of jazz and was frequently referred to as "swing jazz." Jazz was still popular, and as bands became larger many of them emphasized swing jazz. Benny Goodman's band is a case in point, and of course, Duke Ellington's and Paul Whiteman's bands were still at the center of the music world. Swing was danceable music; it was distinguished by a strong rhythm with medium to fast tempos. Other popular swing bands were Count Basie's, Artie Shaw's, Gene Krupa's, and Glenn Miller's. Important musicians of the time were Teddy Wilson and his piano, Lionel Hampton and his vibraphone, Dizzy Gillespie and his trumpet, and of course, the great Louis Armstrong.

The 1940s brought another variation in jazz called bebop. Bebop was quite different from swing, and for the most part it was not danceable music; many referred to it as "musician's music." It was characterized by a very distinct rhythm. Three of the best known early beboppers were Charlie Parker, Dizzy Gillespie, and Thelonius Monk.

In the 1950s there were several other break-offs from traditional jazz. Three of them were free jazz, cool jazz, and later, jazz fusion. All were rooted in bebop and were less structured than previous jazz. Free jazz has a very loose harmony and tempo. Cool jazz is a mixture of bebop and swing, and jazz fusion is a blending of jazz with rock music.

In the 1980s and 1990s there was a resurgence of traditional jazz fueled by Wynton Marsalis, Harry Connick Jr., and others. Marsalis is a well-known New Orleans trumpeter; Connick is a pianist and singer who had his roots in New Orleans.

So, let's go back to our question: how do we define jazz, and what exactly characterizes it? From a simple point of view it is very rhythmic music that borrows heavily from the blues: it uses blue notes, syncopation, anticipation, triplets, and so on. But two things that are very characteristic of it are improvisation and the use of superchords. Improvisation is a major part of jazz, and much of jazz is improvised on the spot. In addition, jazz wouldn't be jazz without the use of substitute or superchords; they sometimes give a slightly dissonant sound to the music, but they are loved by all jazz musicians.

Country Music

Country music is currently one of the more popular types of music; in the United States it has a relatively large audience, and there are more country radio stations across the nation than any other type, including pop. It had its roots in early folk and gospel music, and what is sometimes referred to as "old time" music. The earliest country music was sometimes called "hillbilly music," but by the early 1940s the name was changed to "country music" because "hillbilly music" was seen by many as a degrading label.

Much of country music can be traced to two influences: that of Jimmie Rodgers and that of the Carter family. Jimmie Rodgers is usually considered to be the first country superstar. He wrote songs about the problems of ordinary people—lost love, women, drinking, and life and death. Even though his musical career lasted only six years—he died of TB at the age of 36—his influence was enormous. His rise to stardom began in 1927, when he was "discovered" by talent scout Ralph Peer. Two of his more popular songs were "Blue Yodel" and "I'm in the Jailhouse Now."

The Carter family was also discovered by Ralph Peer, at almost exactly the same time that he discovered Jimmie Rodgers. The group consisted of A. P. Carter, his wife, Sara, and his sister-in law Maybelle. A. P. was the leader of the group, Sara was the lead singer, and Maybelle played the guitar. A.P. was the inspiration for the group; he traveled widely through the hill country around his hometown of Mace Springs, Virginia, gathering songs from the people of the area; he also wrote many songs. Sara and Maybelle arranged the songs; Sara had a beautiful singing voice, and Maybelle had a unique style of playing the guitar. Many later singers were influenced by Sara; these included Patsy Cline, Kitty Wells, Loretta Lynn, Dolly Parton, and June Carter Cash, to mention only a few.

Jimmie Rodgers's music was written for the working man and was about things he could relate to: love, troubles, and so on. The Carter family's music was generally traditional folk music. Most of the country music that came after these two groups followed the style of one or the other. The best known singer of the Jimmie Rodgers branch was Hank Williams. He wrote about love, heartbreak, and unhappiness, but some of his songs—such as "Jambalaya" and "Hey, Good Lookin' "—were quite upbeat. Two of his better-known songs were "Your Cheating Heart" and "I Can't Help It (If I'm Still in Love with You)." His music is still heard extensively today, and his influence was great, despite his short life; he died at the age of 29.

Country music today has many styles. Some of the main ones are

- the Nashville sound
- western swing
- traditional western (Roy Rogers, Gene Autry)
- the Bakersfield sound
- honky tonk
- outlaw country
- bluegrass
- country rock
- rockabilly

The largest genre within country music today is the Nashville sound. It started about 1960 and soon took over country music. Producers such as Chet Atkins made it into one of the most commercially successful types of music on the market. It featured smooth vocals, strings, and background singers. Some of the leading artists of the sound were Jim Reeves, George Jones, Patsy Cline, and Tammy Wynette. About the same time pianist Floyd Cramer developed his "slip note" style of piano playing, and it soon became the standard in Nashville.

The Grand Ole Opry in Nashville played an important role in popularizing the Nashville sound. Almost all well-known country singers have performed on the stage of the Grand Ole Opry. A few of note have been Johnny Cash, Eddie Arnold, George Jones, and Kris Kristofferson.

In the early 1970s a number of musicians broke away from Nashville and gave us the "outlaw" sound. Among them were Waylon Jennings, Willie Nelson, and Kris Kristofferson. And in the 1980s we got the "Bakersfield" sound from Buck Owens and Dwight Yokum.

What are some of the characteristics of country music? There are many left hand or rhythm patterns; most are relatively simple. A typical one is

In the case of the piano the right hand usually plays a more important role. Broken chords are used extensively, as are "walk ups" and "walk downs" as shown.

Of particular importance, however, is Floyd Cramer's hammer and drone. The drone note is a note placed above the melody note; it is usually the root note or sometimes the fifth. This drone note is always the same for a given phrase.

A hammer is a "slip note"; it is like a crushed grace note and is written as follows.

The drone and the hammer have been a staple of country music ever since it was introduced.

The above examples refer to the piano or keyboard, but they also apply to the guitar, which is, of course, the major instrument of country music.

New Age

New Age music is quite distinct from the genres discussed above. Rhythm is not a major characteristic of this music; in fact, for the most part it has very little rhythm. Or at least rhythm is not emphasized. Nevertheless, New Age music has a sizeable audience. It is a relatively recent style, although it is based to a large degree on music that goes back centuries.

The term "New Age" refers to more than music; it is also a lifestyle, and many book and music stores cater to it. A major part of the audience for this type of music are people who adhere to New Age beliefs, but this is far from the entire audience. You merely have to go to a George Winston and Yanni concert to see this.

To a large degree, New Age music is electronic—in other words, based on electronic instruments. But other instruments are also important, particularly the acoustical piano, and indeed, some rather strange instruments and sounds are occasionally used. For the most part, it is instrumental music. It is usually quite melodic but not danceable; furthermore, it is usually very repetitive.

Some people refer to New Age music as "nature music" or sometimes as "waterfall music." And indeed sounds from nature are occasionally used. It is frequently associated with relaxation and meditation, so it is soothing and simple; popular themes are the universe, space, nature, the environment, well-being, and being in harmony with the world.

Some of the better-known artists are George Winston, David Lanz, Yanni, Liz Story, and Jim Brickman. Winston's CD "December" has sold millions of copies, and Yanni has had a tremendously successful career both on TV and in concerts. Interestingly, some of the more successful New Age artists don't like to be tagged with the New Age label. Brickman frequently refers to his music as "adult contemporary," and Lanz, as "contemporary instrumental."

The characteristics of New Age music are

- slow to medium tempos
- very repetitive melodies
- frequent use of arpeggios in the left hand if played on the piano
- primarily diatonic harmonies
- frequent use of pentatonic scales

- frequent use of sevenths, ninths, and elevenths
- occasional use of sounds from nature

Typical left-hand patterns on the piano are as follows:

Longer patterns of this type are also frequently used. Finally, cluster chords such as those below are frequently used in the right hand.

Pop

"Pop music" is a catchall term defining several different types of music. It is characterized by a danceable beat, a simple melody, and a repetitive structure. Rock and roll, rhythm and blues, hip hop, and several other types of music frequently end up on the "pop" charts. One of the major characteristics of pop music is that it is very commercial, primarily a result of CD (and earlier, record) and DVD sales. The pop market is strongly influenced by teenagers, who are the primary market for the CDs and DVDs.

Pop music has, of course, changed significantly over the years. In the 1940s and '50s pop music consisted of the songs sung by such artists as Bing Crosby, Frank Sinatra, Dean Martin, and Peggy Lee. For the most part, the music was relatively smooth, and the "beat" wasn't emphasized. Then in the 1950s and '60s came Elvis Presley, the Rolling Stones, and the Beatles, and pop music changed significantly. With rock and roll, a heavy beat became a staple of the music. A number of other, non–rock and roll artists were also popular during this period; these included the Beach Boys, the Supremes, Neil Diamond, Burt Bacharach, Ray Charles, and Stevie Wonder.

Then in the 1970s came Billy Joel, Elton John, the Jackson Five, the Carpenters, Olivia Newton-John, and others. This was the "disco era." During the 1980s the biggest hit was Michael Jackson's "Thriller," but Madonna also gained considerable popularity during this time. Finally, in the 1990s came Brandy, Selina, Celine Dion, Sheryl Crow, Eric Clapton, and Jewel, and in the 2000s, Britney Spears, Ricky Martin, and Jessica Simpson.

Rhythm and Blues

Rhythm and blues (which is frequently referred to as R&B) combines jazz, gospel, and the blues. Begun in the 1940s, it is a rocking style of music that frequently uses a boogie type base with a strong backbeat. Like jazz, it is usually based on 12 bars, and in many ways it was the predecessor to rock and roll. In addition to jazz, it was strongly influenced by jump blues and gospel music. Most early R&B musicians were, in fact, jazz musicians.

Some of the early jazz bands, such as those of Count Basie and Lionel Hampton, were primarily R&B bands. Early R&B hits were "Blueberry Hill" and "Ain't That a Shame" by Fats Domino, and "Whole Lotta Shakin' Goin' On" by Jerry Lee Lewis (basically R&B, but it hit the charts in several categories).

In the 1960s much of R&B became gospel-oriented and was sometimes given the name "soul music." Artists who recorded it were James Brown, Ray Charles, and Sam Cooke.

Gospel

"Gospel" refers to music that first came out of early African American churches. Some of the first performers were Mahalia Jackson and Sister Rosetta Tharpe. The Carter family also recorded many gospel songs. Basically, gospel is church, or religious, music. There are several divisions within it, including black gospel, southern gospel, Christian country gospel, and contemporary Christian music. Much of today's contemporary Christian music draws on pop and rock and roll. The lyrics are concerned with praising Christ.

Gospel is usually divided into slow gospel and fast gospel. Older standards such as "Amazing Grace," "Just a Closer Walk with Thee," "The Old Rugged Cross," and "Rock of Ages" are usually considered to be slow gospel songs. Fast gospel is generally played much faster and uses many of the devices of rock and R&B. Left-hand patterns such as the one shown below, which is similar to those used in rock and roll, are used extensively:

The right hand (on the piano) is also very rhythmic.

Reggae

Reggae came from Jamaica in the late 1960s. It is characterized by regular "chops" in the backbeat, referred to as "shank." An early, very fast form of reggae was referred to as *ska*. The primary instruments used in reggae are the guitar, drums, and bass; sometimes the keyboard is also used. The music is usually played in 4/4 time and harmonically is relatively simple. Its major characteristic is a strong backbeat.

A simple reggae rhythm is as follows:

Latin American and Hawaiian

With the dance craze on TV, many people have been introduced to Latin dances such as the rumba and tango. Latin American music is very rhythmic and also very danceable. In addition to the rumba and tango, some of the better-known styles are the bolero and calypso. The rhythms of the first three of these is illustrated below.

In the tango the left hand is completely in charge of the beat. It generally consists of a bass note and broken chords.

The term "Hawaiian music" refers to an array of music from Hawaii, including both folk music and modern rock. Hawaiian folk music is largely religious in nature and frequently includes chanting, but it also includes the Hawaiian dance music of the hula. The hula, which is usually performed by hip-swinging girls, is one of the most popular Hawaiian dances.

A large number of Hawaiian songs have become popular over the years. A few of them are "The Hawaiian Wedding Song," "Blue Hawaii," "Sweet Leilani," "My Little Grass Shack," "On the Beach at Waikiki," "Tiny Bubbles," and "Beyond the Reef." Most of them are fairly traditional Hawaiian songs with simple melodies and rhythms, accompanied by the Hawaiian guitar.

Guitars were brought to Hawaii by early European sailors and missionaries and soon became one of the islands' main instruments. Steel guitars were particularly popular; about 1900 Joseph Kekuku began sliding a piece of steel over the strings of a guitar whose strings had been tuned down; this was the beginning of the Hawaiian guitar. A major Hawaiian guitar style is a finger picking style called slack-key guitar; it is named for the fact that the strings are relatively loose.

Recorded Hawaiian music found an audience in America as early as 1915, and the era from 1930 to 1960 is sometimes referred to as the "Golden Age of Hawaiian music." Big bands and orchestras in America were all playing it. One of the most popular singers in Hawaii was Don Ho, with his highly successful hit "Tiny Bubbles."

Classical Music

Classical music was not left to the end because it is the least important. It is very important, and all music owes something to it. It doesn't have the audience of many of the types of music I have mentioned above, but it still has a sizeable one.

Classical music encompasses most of the music from early medieval times through to the early 1900s. It was at its height from about 1550 to 1900. Although many people look upon it as "highbrow" music today because it is usually associated with fine arts and culture, it was the "popular music" of its day. It is distinct from most of the music of today in that it is always written down, although many early pianists were excellent improvisers. Today, however, few classical pianists spend time improvising. Classical music is generally formal in style, and in many cases it is quite technically difficult, much more so than most other types of music.

The term "classical music" was not used throughout much of the history of the music it now applies to; it came into use for the first time in the early nineteenth century. The history of classical music is actually divided into several chronological periods, summarized in table 11.

Classical music is associated with several important musical forms. One of the most important is the *sonata*, a multimovement work writ-

Table 11. The classical music periods

Period	Approximate dates	Description	Representative composers
Medieval	Before 1450		
Renaissance	1450–1600	Increased use of instruments; bass instruments first used	
Baroque	1600–1750	Use of elaborate ornamentation and new instrumental techniques. Harpsichord still being used, but piano starting to replace it	Vivaldi, J. S. Bach, Scarlatti, Handel
Classical	1750–1820	More emphasis on simple melodies with accompaniment; well-defined composition forms (e.g., sonatas)	C. P. E. Bach, Haydn, Mozart
Romantic	1820–1900	Increased attention to melody and rhythm	Beethoven, Paganini, Schubert, Chopin, R. Schumann, Liszt, R. Wagner, Brahms, Tchaikovsky, Grieg, Rachmaninoff, Debussy
Modern	1900–2000	Experimentation with new sounds	Prokofiev, Copland, Stravinsky, Sibelius, Schoenberg, Bernstein

ten for one or several instruments such as the piano, or the piano and the violin. Many sonatas have also been written for small string ensembles and even orchestras. The standard number of movements is four, in which the first movement is fast, the second slow, the third a minuet (a dance), and the last, relatively fast. Another important musical form is the *concerto* for piano and orchestra (and for other instruments and orchestra); it also consists of several movements. In addition, there are numerous symphonies for symphony orchestras, and all of the better-known composers in the classical period wrote operas. There are, of course, many other types of shorter pieces—nocturnes, ballades, preludes, etudes, and impromptus.

The baroque era is best known for elaborate musical ornamentation, the development of new instrumental playing techniques, and changes in musical notation. The idea that music should represent the emotions of real life was introduced during this era, and this led to music that was designed to move the listener emotionally. Opera was also established as a musical genre during this era, and as you might expect, the early years of opera (and also of the baroque era) were dominated by Italians. Antonio Vivaldi and Domenico Scarlatti were both Italians. Later, however, the baroque era was dominated by Germans. Two of the greatest composers, J. S. Bach and George Frideric Handel, were German. Handel is most famous for his oratorio *Messiah*.

Following the baroque era came the classical era. One of the most famous musicians of this period was Mozart. It's hard to believe that anyone could do so much in such a short time (he died at the age of 35). A child prodigy, he was composing songs at the age of five and played the violin and the keyboard (usually the harpsichord; the piano was still by no means a universal keyboard instrument) in concerts throughout Europe before he was a teenager. His sonatas are still some of his best-loved pieces. In addition, he wrote numerous symphonies, concertos, operas, and quartets. Some of his better known operas are *Don Giovanni*, *The Marriage of Figaro*, and *The Magic Flute*. His music is still played extensively in concerts today.

Another early great of this era was Beethoven, who wrote nine symphonies, five piano concertos, numerous pieces for chamber orchestra, and 32 piano sonatas as well as many other piano compositions and compositions for instruments with piano. In his Third Symphony, the "Eroica," his creativity as a composer is clearly evident. It is still a favorite with modern audiences, along with the Fifth Symphony. It's amazing that even after he began to go deaf he still managed to write seven symphonies, three piano concertos, several piano sonatas, an opera, and other music. Among his sonatas is the beautiful "Moonlight Sonata" and the "Appassionata Sonata." In many ways he changed music—transforming the sonata, the symphony, and the concerto into "grandiose" forms.

In the romantic period, Chopin also wrote a large amount of music, mostly for the piano. At 8 he was hailed as the new Mozart, but by 18 he had suffered a nervous breakdown and was troubled by periodic ill health from then until his death at the age of 39. Some of his more famous piano pieces are the Polonaise in A flat, the "Polonaise Militaire," the "Fantasy Impromptu," and the Nocturne in E flat.

Another composer of this period, Franz Liszt, was a sharp contrast with Chopin in that he was flamboyant and the ultimate "showman": he was "the pianist" of his time, one of the greatest virtuosos of all time, and he had a large impact on music. Liszt was a prolific composer, but much of his music was too difficult for the musicians of the time. His Hungarian Rhapsodies, for example, generally considered to be "show-off" pieces, were so difficult that hardly anyone other than Liszt could play them. He also wrote many beautiful etudes. His most familiar composition may be the dreamy nocturnes entitled "Liebesträume."

The romantic ended at the beginning of the twentieth century and was followed by the modern era, which introduced a radically different type of music. Composers began experimenting with new musical forms that incorporated irregular rhythms, new scales, atonalism (not having a definite key), and impressionism. This led to sounds that went against the grain of the time, and many people found it difficult to listen to. Gradually, however, it was accepted. Early composers of this period such as Debussy (who overlapped the romantic and modern periods) and Richard Strauss were much less radical than later composers such as Stravinsky and Prokofiev.

One of classical music's major features is its durability. Many of the classical pieces that are still played today are hundreds of years old. Most musicians are classically trained in their early years; they may go on to other types of music, but it is accepted that classical study early in life is needed by all serious musicians.

MUSICAL INSTRUMENTS

III

8

Why a Piano Is Not a Harpsichord

Rubinstein lowers his hands as he finishes a Beethoven sonata. The audience breaks out in applause. As he rises from his bench and bows, the applause increases, then several people begin to cheer. He bows again and begins to leave the stage. The applause and cheering swell as he disappears behind the curtain. It continues for several seconds, then suddenly he reappears and walks quickly to the piano. As he sits down the hall goes quiet; then he plays the first few bars of a Chopin polonaise, and a roar of delight resounds throughout the crowd.

I've had the pleasure of hearing Rubinstein and other classical pianists such as Van Cliburn and José Iturbi, as well as many popular pianists such as Peter Nero, Liberace, and Roger Williams. Roger Williams, in fact, got a degree from the university I taught at for many years. Everyone who knows anything about Williams knows that he studied at Juilliard, but few know that he also got a degree from Idaho State University. Interestingly, he also majored in music at Drake University but was expelled for playing "Smoke Gets in Your Eyes" in the practice room.

The piano is truly a magnificent instrument: ask any pianist. I have to admit that I played it for many years before I really knew how it worked. It was easy to see the hammers striking the strings, but the sound that came from them seemed amazingly loud, and I wasn't sure why. I eventually found out that most of the sound wasn't coming from the strings; it was coming from a large soundboard. To hear a musical sound you have to get a large amount of air vibrating, and a steel string can't do that. If the string is attached to a soundboard, however, the string transfers the sound to the soundboard, which amplifies it. In short, the soundboard sets a large amount of air vibrating, and it is this vibrating air that impinges on your eardrums.

One of the main reasons the piano is so popular is that it is one of the few instruments on which you can play both the melody and the harmony. In this respect it is like an orchestra.

The Beginnings

If you look inside a piano, you see that it is a relatively complex instrument. It has thousands of moving parts, along with hundreds of strings. The first pianos didn't have all these parts, and indeed, the piano has evolved over many years. Its origins go back hundreds of years to much simpler stringed instruments. Two of the earliest of these instruments were the psaltery and the dulcimer. The psaltery is so old, in fact, that it was mentioned in the Bible. It consisted of several strings stretched across a frame or hollowed-out gourd, and it was played by plucking the strings, as can be seen in figure 66. The dulcimer, which originated in the twelfth century, was similar to the psaltery, but the strings were struck with small wooden hammers.

Aside from the strings there is, of course, little resemblance between the psaltery or dulcimer and the modern piano. The addition of keys was the first significant step in the evolution from a simple stringed instrument to the complex instrument we know today. Keys were actually used first on early organs; they were added to stringed instruments in the fifteenth century. One of the first to use keys was the clavichord. It grew out of Pythagoras's monochord, which was used to study the relationship between vibrations on a string.

Fig. 66. Girl playing a psaltery.

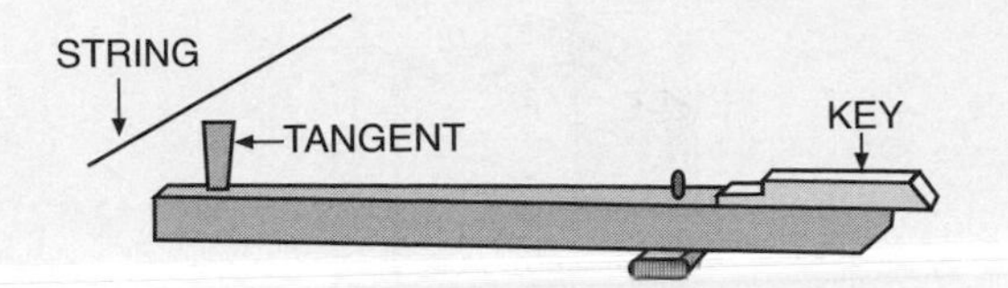

Fig. 67. Mechanism for striking the string of a clavichord.

The clavichord had about twenty strings which were caused to vibrate by pushing a bridge, or "tangent," against them (fig. 67). A given string could be struck at several points, so several different vibrational modes could be initiated in a single string. A damper was used on the shorter section of the string so that it would not vibrate. In at least two respects the clavichord was like the piano: its strings were metal and it had a soundboard (which was not attached to the frame).

The clavichord was a favorite family instrument for many years, being frequently found in the homes of the time, but the instrument had a serious problem. You could vary the loudness slightly, depending on how hard you struck the keys, but the sound that emanated from the strings was not loud, and therefore clavichords were not suited for public performance.

The Harpsichord

One way of making the sound slightly louder was to pluck the strings, as in the psaltery. The earliest records of keyed instruments of this type date to about 1400. In these instruments the key was connected to a wooden rod called a jack. Set into the jack was a tongue that carried a plectrum or quill set at right angles to the tongue (see fig. 68). When you pressed the key, the jack rose and the quill plucked the string, then fell back. Through an ingenious hinge, the tongue moved back so that the string didn't strike the quill on the way down.

Instruments called virginals were the first to use this device. They were small and rectangular, with one string per note running parallel to the keyboard. Another instrument called a spinet came out about the same time; it was similar but had its strings set at an angle to the keyboard. The strings, in this case, were arranged in pairs, with the jacks in the spaces between the pairs. Virginals and spinets were early forms of what we now call harpsichords, but when we talk about harpsichords today we are usually referring to the more advanced models that are shaped like a grand piano.

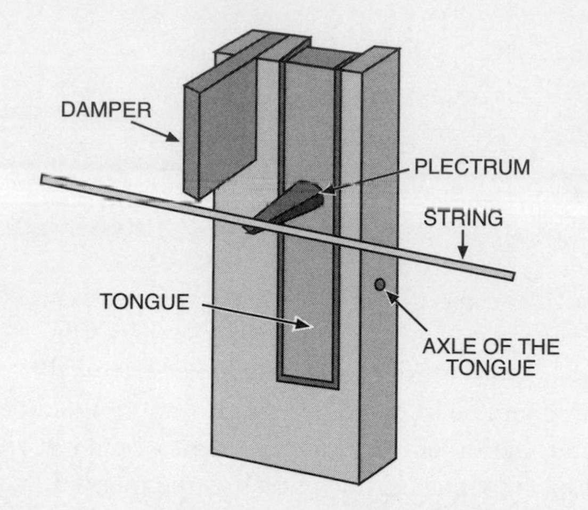

Fig. 68. Mechanism for plucking the string of a harpsichord.

As inventors strived to improve the early harpsichords, strings got longer and were under more tension, and soundboards got larger. Eventually the strings were so long that they were placed perpendicular to the keyboard. This caused a change in the shape of the instrument. The treble strings were much shorter than those in the bass, and the case began to follow the length of the strings, giving a wing shape, similar to that of the grand piano (fig. 69). Foot pedals were also added to damp the strings.

The first harpsichords were built in Italy. They were of light construction and had relatively low string tensions; as a result, their tone was very soft. Around 1580, however, Hans Ruckers of Flanders began building much sturdier instruments. He used longer strings and put them under greater tension, and he made the cases heavier and introduced a spruce soundboard. For several years his harpsichords set the standard, but soon harpsichords were also being made in France, Germany, and England. The French introduced more keys (for a total of about five octaves), and it was possible to vary the combination of strings being struck in French instruments. In addition, both the Germans and the English made advances that improved the design of the harpsichord.

The harpsichord was an extremely popular instrument for a number of years, but after the piano was invented, the harpsichord gradually fell out of favor. Interestingly, it has made a comeback recently. One of the people who was important in this renaissance of

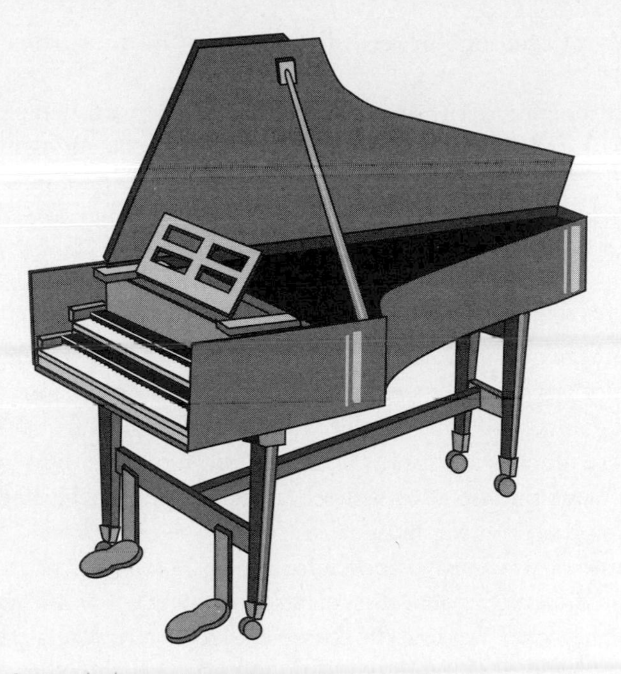

Fig. 69. An early harpsichord.

the harpsichord was Wanda Landowska of Poland; she is generally considered to be the greatest modern harpsichordist. In fact, several harpsichord makers actually made harpsichords just for her. She particularly enjoyed Bach, and once said to another concert artist, "You play Bach your way, and I'll play him *his* way." Bach, of course, composed most of his keyboard music on the harpsichord, but today it is played mainly on the piano. Landowska wanted people to hear what this music sounded like when Bach composed it, and indeed, she had a very successful concert and recording career.

More recently, the harpsichord has been used in the recording sessions of several popular artists. Both the Beatles and the Beach Boys used harpsichords in the background in some of their recordings. Jimi Hendrix also used it, as did Linda Ronstadt in her song "Long, Long Time." The harpsichord is also occasionally seen on TV. It was frequently used on the *Lawrence Welk Show* and can be seen in reruns of the show; furthermore, it was used in the background in the 1970s TV show *The Partridge Family*. Surprisingly, Tori Amos also used one on *Saturday Night Live*. And very recently,

Paula Abdul used one in recording her hit "Blowing Kisses in the Wind."

With their modern revival, harpsichords are still manufactured. Hubbard Harpsichords of Framingham, Massachusetts, still makes them (mostly in the form of kits). The reason that there are few around today—compared with the very large number of pianos—is, of course, their lack of dynamics. You can't vary the sound.

The Pianoforte, or Piano

How could the harpsichord be improved to give better dynamics? You could lengthen the strings and put them under more tension. This was done, of course, and it helps, but unfortunately there are limits when it comes to plucked strings. As it turned out, the solution to improved dynamics was the use of an entirely different approach. The harpsichord was based on the psaltery, in which the strings were plucked. But there was also another model for the action of a keyboard instrument: the dulcimer, where the strings were struck by a hammer. The problem, however, was that the strings had to be struck relatively hard, and the hammer had to get out of the way fast. In the case of middle C, for example, it had to jump back in less than 0.0038 seconds.

The first to tackle the problem was Bartolomeo Cristofori (sometimes spelled Christofori) of Padua, Italy. Born in 1655, Cristofori was both a musician and an inventor. When he was 33 years old his work came to the attention of Prince Ferdinando de Medici of Florence (the same family that supported Michelangelo). Ferdinando was a music lover with a large number of valuable musical instruments, and he needed someone to take care of them. He was also interested in improving the harpsichord, and this is no doubt what attracted him to Cristofori, who was already well known as an inventor. Cristofori was reluctant at first to leave Padua, but Ferdinando made him an offer he couldn't refuse. He was to have his own house, a shop, several assistants, and a good wage.

Cristofori began working on what would eventually become the piano in about 1700, but it wasn't until 1709 that he introduced it to the public. It was a relatively crude instrument at this stage, and while it used hammers to strike the strings, the hammers had little speed and therefore didn't get out of the way of the vibrating string fast enough. Cristofori continued working on his new instrument, however, and by 1720 he had come up with an ingenious way of speeding up the hammers. He used a small "catapult" that gave the hammers three times the speed they previously had. He also used heavier strings and put them under greater tension. Furthermore, the hammers now struck two strings rather than one. The hammer heads were made of compressed paper in an effort to keep them light, and they were surrounded by leather, and finally, the soundboard was supported by a frame to keep it from warping.

Cristofori called his new instrument the pianoforte, which means "soft-loud." He chose this name because, unlike the harpsichord, the pianoforte could be played both softly and loudly. You might think that because of this advantage in dynamic variation, the pianoforte would quickly eclipse the harpsichord, but it didn't. A lot of improvements had been made to the harpsichord, and the first pianos were barely louder than the harpsichord. In fact, they sounded much like the harpsichord.

Bach didn't like the first pianos. (I'll call them pianos from now on.) He thought they were stiff and hard to play. But Cristofori and others continued to work on the mechanism and other parts of the instrument. Heavier strings were used to increase the loudness; heavier cases were developed so that the strings could be put under greater tension; and the mechanism associated with the hammers and keys was improved. And over time the piano gradually became easier to play, and more popular.

One of the major problems with the first pianos was that they were very expensive. Only the rich could afford them. But as time passed, their price came down, and more and more musicians turned to them, delighted with their dynamics. Bach eventually bought one, and although most of Wolfgang Mozart's early work was composed for the harpsichord, he also soon turned to the piano. As a child he played the harpsichord, but on a trip to Paris in 1778 (when he was 21), he encountered the piano, and there was no turning back. From that time on he played the piano, and all his music was composed for the piano. It is said, in fact, that he got so fascinated with the mechanism of the piano, he experimented with trying to improve it.

The piano was well established by the time Beethoven came along. Early on he used Viennese pianos but was never happy with them (their keyboard had only five octaves). He urged piano makers to make pianos stronger and more sonorous, and when John Broadbent of London shipped him a grand piano with six octaves that was much more rugged than the Viennese pianos he was ecstatic. For years it was his pride and joy, but as in the case of his other pianos, he never looked after it. Someone visiting him late in his life reported that the piano was in terrible shape, with several broken strings and numerous stains on it from spilled drinks.

Then came Chopin and Liszt. No one can say Chopin was hard on pianos. His playing was so delicate that some people complained he played too softly. Although he was certainly one of the greatest pianists of his time, he was also one of the least flamboyant. One of the reasons, no doubt, was that he was sick during the last years of his life. While in Paris he had two pianos: a Viennese Pleyel and a small cottage piano he used to accompany his students. He is said to have preferred a piano with a light touch.

In many ways, Liszt was the opposite of Chopin. He was a flamboyant performer who lifted his hands from the piano and brought them crashing down on the keys, so, as you might expect, he was hard on pianos. And according to Chopin, he was. "He would get so emotional when he played," said Chopin, "he smashed several of his pianos."

The piano remained basically the same until the mid-1800s, when the German American piano maker Heinrich Steinway introduced the first cast iron frame. It could withstand much more tension from the strings, and it revolutionized the design of the piano, giving it unprecedented power and brilliance. Steinway had come to America in 1851 with his wife and three sons, and within a couple of years they had begun making pianos in New York City. Soon he was overwhelmed with orders, and his business flourished. The Steinway piano soon became the standard. Figure 70 illustrates the now-familiar grand piano, manufactured by many companies.

The Inner Workings of the Piano

One of the most important parts of a piano is the strings. When it is struck, a string vibrates with a frequency given by

$$f = 1/2L\sqrt{(T/\rho)},$$

where

L = length of the string
T = tension in the string
ρ = density of the string

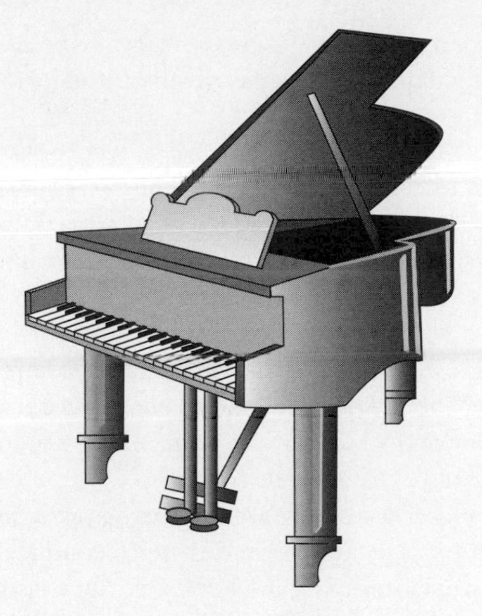

Fig. 70. A grand piano.

Assuming the density to be 0.0059 (the density of steel) we can easily calculate the tension on a given string using this formula. Let's do it for middle C, which has a frequency of 262 Hz and a length of 0.82 m. We get 1,051.7 newtons. Converting newtons to pounds (1 N = 0.225 lbs) we get 236.6 lbs, which is a relatively large force. Later we will see that there are approximately 226 individual strings on the piano (this number varies somewhat, depending on the piano). The force on each of them differs slightly, but for convenience we will take it as 230 lbs; the total force on the frame holding them is therefore 51,980 lbs—a very large force. This is only approximate, but it makes the point that there is a lot of tension on the frame holding the strings, and this is, of course, why frames are made of cast iron.

Looking at the formula, we see that longer strings need a greater tension for the same frequency. But the formula also shows us that we could make the string shorter if we decreased the tension. Why isn't this done? As it turns out, there are several reasons for long strings and high tension. First of all, it's important that the string doesn't bend too much; if it does, the forces of rigidity within the metal string begin to interfere with the vibrations. Longer strings are therefore better, as they bend less. Second, we want the kinetic energy of the string to be as high as possible, as it is related to the loudness of the sound. Kinetic energy is given by the formula

$$E = \frac{1}{2} mv^2,$$

where m is mass and v is velocity. Both the mass and velocity of the string depend on its length for a given thickness. This is easy to see in the case of mass (a longer string is heavier), but in the case of velocity we have to look at the geometry of the string. We know that the stiffness of a wire affects the vibration of a short one more than it does a long one. This means that the greater the length, the greater the velocity, and the greater the kinetic energy. This, in turn, translates to a greater loudness. So it is in our interest to make the string as long as possible.

Let's take a closer look now at how the hammer hits the strings. The mechanism on a modern piano is quite complex, consisting of many parts, so we will simplify it by considering Cristofori's original design. From the diagram in figure 71 we see that as you press a key, a jack hits what is called the intermediate lever. The end of this lever kicks the stem of the hammer, catapulting it against the underside of the string. At the same time a damper above the string is raised so that the string is free to vibrate. When the hammer falls back, it is caught by a "back check" so that it doesn't rebound and hit the string again. This mechanism allows for rapid repetition of notes, which is important in many piano pieces. When your finger is still on the key, the intermediate lever is still raised and ready to be catapulted again. When the key returns to its original position, the damper comes back to rest on the string.

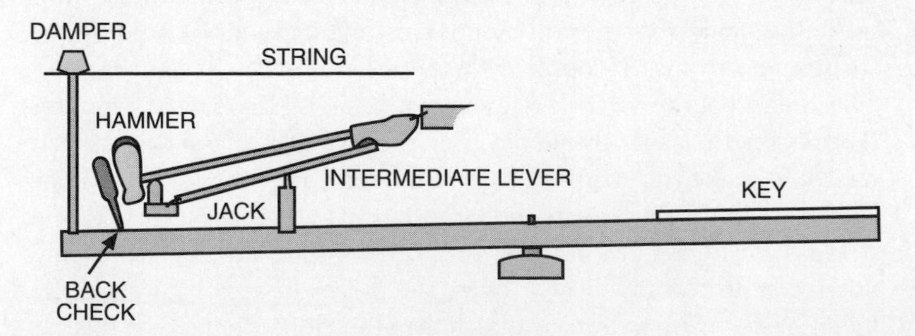

Fig. 71. Mechanism for striking a string in a piano.

In Cristofori's original piano the outer layer of the hammers was leather, but leather was eventually found to be unsatisfactory. They are now made of felt. In grand pianos the hammers fall back due to gravity; in uprights they bounce back after they hit the strings. The more efficient action of falling is one of the reasons grand pianos have a smoother action than uprights.

The sound is created by the vibration of the string, but after a very brief interval called the attack time, the vibration is transferred via a wooden bridge to the soundboard. From here it is radiated to the air. The soundboard is about 1 cm in thickness and is usually made of a low-density, elastic wood such as spruce.

Finally, most pianos have three pedals (the center one is sometimes missing). The left pedal is the soft pedal. If this pedal is depressed in a grand piano, the entire mechanism (or action) is shifted slightly so that when you strike a note, the hammer hits two strings instead of three, decreasing the loudness. (In upright pianos, the soft pedal works by moving the hammers closer to the strings, so that they hit the strings with less force.) The right pedal is called the forte, or sustaining, pedal; it lifts the dampers from all the strings so that they are free to vibrate. The tone from the note will continue to sound until it dies off naturally, but when the sustaining pedal is released, the tone is damped immediately. The center, or sostenuto, pedal, keeps the dampers raised only from the strings that had been struck when the pedal was depressed.

Magical Overtones

Consider the note G_4 (in the fourth octave). We know that when we strike it, the G string vibrates with a frequency of 396 Hz. But when you play the same note on a violin, it sounds slightly different, even though it has the same frequency (fig. 72). In fact, if you set the frequency of a signal generator at 396 Hz, the tone will sound different again. The reason for this can be seen if we display the waveforms on an oscilloscope; although each is indeed 396 Hz, the shape of their envelopes is different. In other words, each note has a characteristic shape; we refer to this as its *timbre*.

What is the reason for this? The answer can be found by looking closely at the vibrating strings. In chapter 4 we talked about overtones; we saw that if you struck a string, you would excite its fundamental frequency, but you would also excite several overtones. These

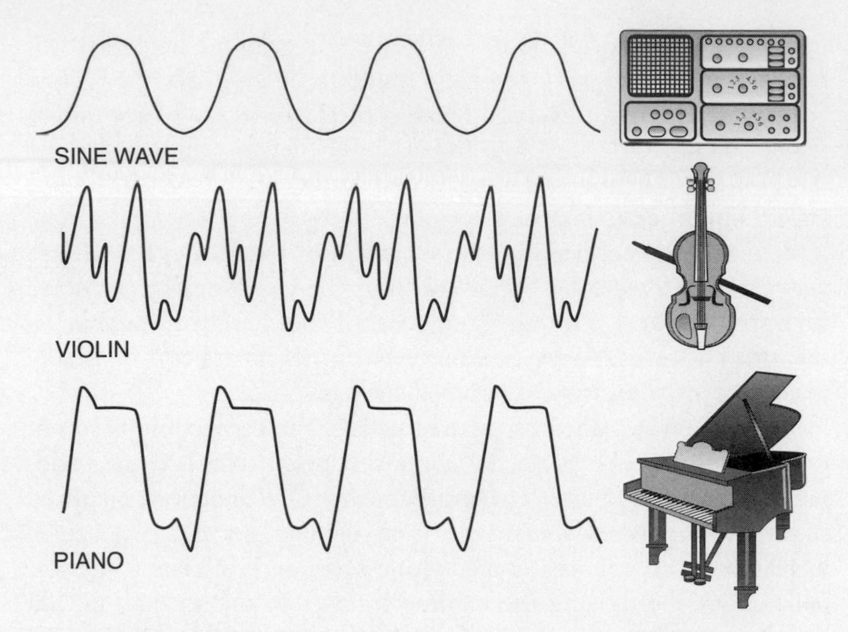

Fig. 72. The waveform of G_4 as sounded on a signal generator, a violin, and a piano.

overtones are generally fainter than the fundamental. In physics, the frequency of the overtones are whole-number multiples of the fundamental frequency. We refer to this system as being "harmonic." When you strike a key on the piano, you get the fundamental and several overtones, and you might expect that they would be integral multiples of the fundamental—in other words, that they would be harmonic. But because of the stiffness of steel wire, this is not the case. The overtones on a piano become increasingly sharp, or higher in frequency, compared to the overtones of a pure harmonic tone, as the frequency increases. These overtones are referred to as "inharmonic." This is why the timbre of a piano is different from that of a violin. The strings of a violin are made of gut rather than steel and are therefore not as stiff as those on a piano. Because of this, the overtones from a violin are closer to being harmonic than those from a piano; therefore, its timbre is different. And this brings us to another reason for long strings and high tensions. As it turns out, the problem of inharmonicity is minimized by using long strings and keeping the tension as high as possible.

Indeed, a pure tone is rarely heard in music. Although some of

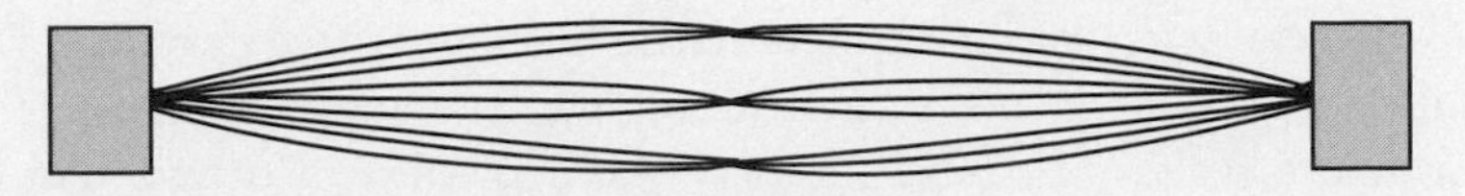

Fig. 73. A vibrating string showing the first overtone.

the woodwinds are close, none is exact. But how do these overtones come about? If you look closely at the string (fig. 73), you see that they are superimposed on the fundamental. The fundamental produces a single loop over the length of the string, and at its two ends are nodes. The first overtone has two loops and is half as long as the fundamental; higher overtones are half as long again. To further complicate things, the overtones have different degrees of loudness, and they don't all decay, or die away, at the same rate.

It might seem that overtones would be a problem and undesirable. But they aren't. It's overtones that give "warmth" to a note. A certain amount of inharmonicity is therefore desirable in the piano. In fact, if a piano were tuned so that was exactly harmonic, it would sound out of tune.

Tuning and Other Tidbits

Before we talk about tuning let's take another look at the keyboard. The modern piano has 88 notes and consists of 7⅓ octaves. It covers a frequency range from 27.5 Hz up to 4,187 Hz. In the modern piano, strings for the lowest bass notes are wrapped in copper (or iron) and are under the lowest tension. The number of wires per note also varies throughout the piano: in a grand piano, the lowest 10 have one wrapped wire, the next 18 notes have two wrapped wires per note, and the upper 60 are unwrapped with three strings per note (these numbers vary slightly from piano to piano). The strings for the highest notes are the thinnest and are under the most tension. Because of the varying thicknesses and tensions of the strings, the timbre is not the same for all notes.

Each string is attached at the keyboard end to a separate tuning pin, which passes down through a hole in the frame and is anchored on a strong wooden pin block. This pin block is built up of numerous hardwoods that are at cross layers. At its other end, the string is attached to what is called a hitch pin.

In order to understand tuning we have to briefly review some of the scales on the piano. As we saw earlier, the white notes make up the diatonic scale, and if we include the black notes we get the chromatic scale. Furthermore, each successive frequency change in the chromatic scale is called a semitone, and finally, an octave has 12 semitones. The white keys of the piano are not tuned exactly to the diatonic scale, or we can say that all the notes (including the black ones) are not tuned exactly to the chromatic scale (as it was defined earlier). Rather, they are tuned to the equally tempered scale in which the octave is divided into 12 equal intervals.

What is the reason for this? The major reason is that if you played a chord on, say, a piano with the regular chromatic scale tuned to C, it would not produce acceptable chords in another scale, for example, G. What this means is that if a piece is written in C and you decided to play it in G, it would sound strange—many of the chords would be dissonant. To get around this, the chromatic scale is divided into 12 equal half tones; in this case the frequency ratio of any two adjacent notes is always the same.

Even when a piano is perfectly tuned, it actually has small dissonances in it, and there is no way of getting rid of them. The problem, of course, is that the chromatic scale is discrete and does not allow frequencies between the notes. But unless you have a particularly good ear, you'll never notice it.

I won't go into the details of tuning; it is a fairly complex and tedious process. It might surprise you, however, that the tuner comes with only a tuning fork, a few rubber wedges, and some felt strips. (Some now use electronic tuners rather than tuning forks.) But tuning is done mostly by ear, in particular, by listening to beats. The tuner usually begins by tuning a particular string—say, middle C—using a middle C tuning fork. Then all the octaves of this note are tuned, since they are approximately integral multiples of it. This is done using a tuning lever or socket wrench to adjust the tuning pins. The frequencies of two notes an octave apart on a well-tuned piano will not be in the ratio of 2:1. This is because of the inharmonicity inherent in a piano. It is called the "stretched" octave. The tuner then proceeds to other strings by listening to the beats created when two strings vibrate simultaneously.

Touch

Piano teachers spend a lot of time telling their students how to press the note down. And you have no doubt heard the difference in technique between various pianists. Good pianists are able to vary the sound considerably. But can you actually change the timbre of the note by the way you strike it? Most piano teachers would say yes, but there is considerable controversy on this subject. When a physicist looks at the action of a piano, he sees that in the fraction of a second before the hammer hits the strings, it has been thrown clear of the mechanism; this mean that the pianist has no control over the note during this time. So the player can change the intensity or loudness of the tone by imparting more or less kinetic energy to the key, but she cannot change the timbre. In short, everything would be the same if the key was depressed mechanically. (We're assuming, of course, that she is not changing the damping, as the damper pedals change the timbre.)

Now let's look at the other side of the argument. The argument here is that the "flex" of the hammer shaft also affects the tone, and the shaft is significantly flexed while it is in free flight, particularly if the note is played loudly. The effect is likely to be greater in the bass because it has the heaviest hammers. And there are indications that the flex does have an effect. I'm not going to hazard a guess as to who is right, however, as it would no doubt generate a lot of controversy.

Conclusion

We now come back to our question: Why is a piano not a harpsichord? Well, they are similar; you'll have to admit that. But the difference in the way the note is struck makes a big difference in the sound that is produced. The string in the piano is struck, and the string in the harpsichord is plucked. It's as simple as that.

9

The Stringed Instruments

Making Music with the Violin and the Guitar

Everyone over a certain age has heard of "Beatle-mania" or "Elvis-mania," the frenzy among fans whenever the Beatles or Elvis Presley stepped onto a stage. And of course, since their time, many other musicians have had the same effect on audiences. It might seem that this is a relatively modern phenomenon, but it isn't. More than 200 years ago the violinist Nicolo Paganini (fig. 74) drove his audiences into a frenzy that had never been seen before. Night after night he left people in awe, so inspired by his playing that many seriously believed he had a pact with the devil. The German poet Jakob Boehme said of Paganini's playing, after attending a concert, "I have never seen its like in my life." The composer Franz Schubert said, "I have just heard angels sing." And pianist Franz Liszt, "What a man! What a violin! What an artist!"

In many ways Paganini did indeed act like the devil. He would arrive at a concert in a black coach drawn by four black horses. Dressed entirely in black, he would rush onto the stage, brush back his long black hair, and begin playing. And unlike other musicians of the time, he never used music; he memorized everything. This was something new to audiences, and it was part of what led to his mystique. How could anyone memorize hours of music as he did? Furthermore, he played passages that other violinists swore were impossible to play:

Fig. 74. Nicolo Paganini.

he is said to have sometimes played more than 12 notes a second and could play the most difficult pieces ever written at sight. And it was not just his technical wizardry that hypnotized the audiences: he could play slower passages so tenderly and beautifully that women in the audiencc broke into tears. He was a first-rate showman, and to top off his showmanship he loved to play tricks on the audience. Occasionally he would bow his violin so forcefully that he would break all the strings on it except the G string (on purpose, of course); then he would continue and finish the concert using only the G string.

But where did he find violin music that could showcase his extraordinary talents? Very little highly technical violin music had been published by that time. To get around this he composed his own music. Not surprisingly, little of it has survived: it was too difficult for other violinists to play. Later in the chapter we will learn a bit more about Paganini's life and career, but first we begin by learning more about the instrument he played, starting with its predecessors.

The First Stringed Instruments

One of the first stringed instruments was the *lyre*, which existed in early Egypt. The first lyres had four to six strings, but eventually the number settled at four; a six-string lyre is shown in figure 75. The strings were stretched from a bar across a U-shaped yoke that was held in the left hand and were strummed with a plectrum held in the right hand. Lyres were used mostly to accompany singers. It wasn't long, however, before musicians began to feel restricted by four strings and added more. Indeed, they added a large number more, which eventually led to the *harp* (which had about 25 strings). The harp was a favorite instrument of David of the Bible.

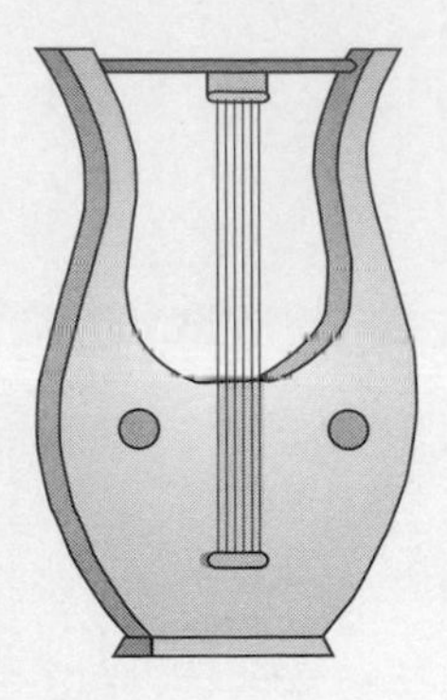

Fig. 75. A lyre.

All early stringed instruments were plucked. It was known that the tone or pitch of a string could be changed by changing the length of the string. The easiest way to do this was to press down on the string with the fingers, but musicians needed to know where to press down to achieve a given tone, and this led to the addition of *frets.* The strings usually passed along a narrow neck, and the frets, which were usually made of strips of animal gut, were placed along the neck. The player could then press the strings down along the edge of the frets.

In the sixteenth century a small instrument called the guitarra, with six strings and four pairs of frets along its fingerboard, became popular in Spain. Strummed with a plectrum, it was used extensively for dancing. It was improved over the years and eventually became the modern guitar, which we will examine later in this chapter.

Another important early instrument was the *lute*, which had six double strings and moveable gut frets. It was very light, and like the lyre, it was used mostly for accompanying singers. During this time the plectrum was discarded, and most musicians began strumming the lute with their fingers.

All early stringed instruments were plucked. It is perhaps strange that it took so long for musicians to discover that "bowed" strings also produced beautiful tones. Interestingly, two types of bowed instruments came on the scene independently about 1500, during the Renaissance. The first were instruments in the *viol* family, which developed in Spain. There were three instruments in this family—a bass, a tenor, and a treble—and each had six strings tuned a fourth apart. The second family was the *violin* family, which developed in Italy. It had four strings and eventually completely overshadowed the viol family.

The Making of a Violin

Basically, a violin consists of a front plate and a back plate, both of which are arched outward slightly, separated by ribs, or sides, to give a hollow "box" called the resonance box. Four strings are stretched from the tail piece across a bridge and over a fingerboard to the nut at the pegboard end (fig. 76).

The strings are under tremendous pressure (the total for the four strings is about 50 lbs), and this creates a downward pressure of about 20 lbs on the face or belly of the violin through the bridge. The wood of the top plate is very thin, usually only a few millimeters thick, so it needs reinforcement. This comes in the form of two supports: a post and a bar. The first is called the *sound post*, and it is placed directly under one edge of the bridge; the second is called the *bass bar*, and it is placed on the underside of the top, along the bass string (fig. 77).

The design of each part of the violin is very specific because the tone depends strongly on what type of wood is used and exactly how the instrument is put together. The top plate is made of soft wood, usually spruce or pine, cut so that the grain is lengthwise along the violin from the pegboard to the tail. The ribs (sides) and back are made of a hardwood such as maple, frequently a type known as curly maple. The neck, pegboard, and scroll are also usually made of maple,

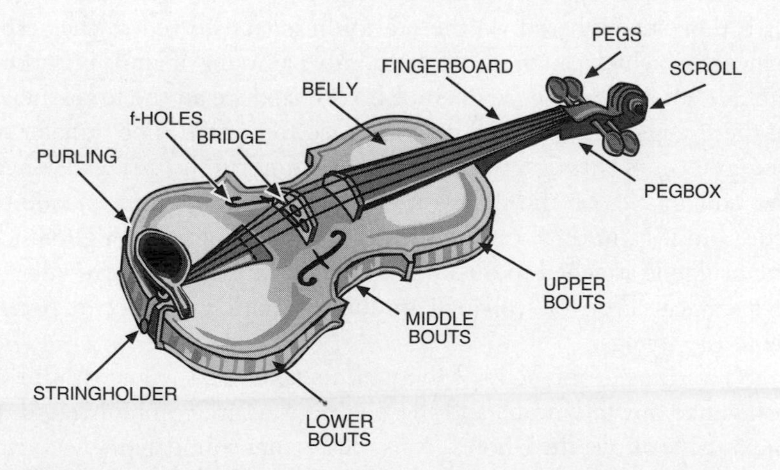

Fig. 76. A violin, showing its parts.

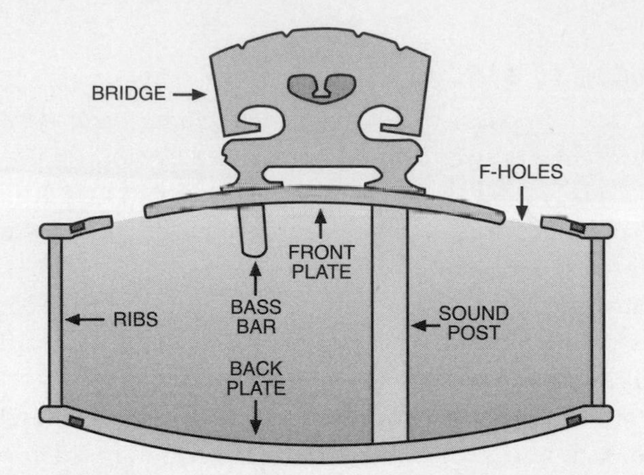

Fig. 77. Cross section of a violin showing the sound post and base bar.

and finally, the pegboard is ebony. After everything is assembled and glued together carefully, the case is varnished to a high gloss.

The Resonance Box

From a simple point of view the violin is merely four strings that are strung across a resonance box that acts to amplify the sound from the strings. Actually, there is no amplification in the technical sense; it's just that the strings don't move enough air for the sound associated with them to be heard, so the vibration is transferred to the resonance box, which can move a considerable amount of air. It is important, of course, that the resonance box reproduce all the frequencies of the strings faithfully, and it is designed to do this. The transfer of energy occurs through the bridge that supports the strings. When the bow pushes on the strings it causes them to move from side to side, and this, in turn, causes a rocking motion of the bridge. Since the bridge is attached to the belly of the violin, it causes it to vibrate in a vertical direction. (A small amount of motion also occurs parallel to the strings.)

Vibrations of the belly of the violin also cause the air within the resonance box to vibrate, and these vibrations are communicated to the outside air via the f-holes. This means that sound actually comes from two sources: the vibrations of the wood and the vibrations of the air within the resonance box. In fact, the amount of sound from each source is about the same. As we saw earlier, a flexible surface, such as that of a violin plate, will vibrate in harmony with any frequency that is applied to its surface. At some frequencies the amplitude of vibration will be large; at others, it will be small. It is, in fact, usually very large at one particular frequency, and we refer to this as the resonant frequency. A vibrating volume of air also has a resonant frequency. This means that we have two resonant frequencies associated with a violin. They are referred to as the main wood resonance (MWR) and the main air resonance (MAR). The MAR depends on the volume of the resonance and the area of the f-holes. It is usually around 280 Hz, which is close to the frequency of one of the four strings. The four strings of a violin are usually tuned to the four notes G_3, D_4, A_4, and E_5; because D_4 has a frequency close to 280 Hz, the MAR strongly reinforces it. The MWR, on the other hand, has a frequency near 420 Hz, which is close to A_4, and the MWR therefore reinforces this A tone.

A particularly useful curve for judging the merits of a violin is called the *loudness curve*. To make such a curve, a violin is bowed to produce its loudest tone across its entire frequency spectrum. As figure 78 shows, the loudness varies considerably from note to note; the MAR and the MWR are approximately coincident with D and A, respectively, and are within the spectrum. A good violin has these two frequencies a fifth apart (and near the notes D and A); if they are more than this interval, the violin is usually judged to be of poor

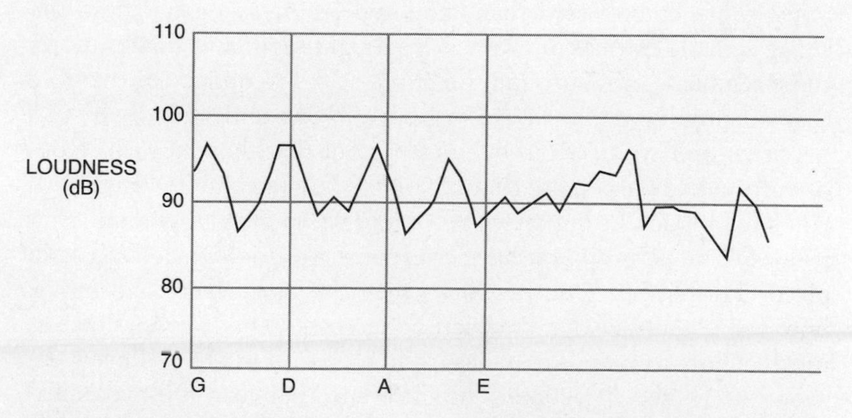

Fig. 78. Frequency spectrum of a violin showing the notes G, D, A, and E.

quality. It is also possible that a strong coupling can occur between the strings and the MWR. When this happens, it causes an undesirable warble called the *wolf tone.* Violin makers try to make sure this tone falls outside the usual tones of the scale.

The Bow and the Strings

The violin's bow is usually strung with horsehair. Like all natural types of hair, horsehair has frictional properties that are quite different in one direction than in the other. You merely have to pull your fingers along a strand of your own hair to see this. This difference in friction is a result of tiny overlapping scales along the hair that are all aligned in the same direction. In a violin bow, half the hairs are aligned in one direction, and half in the other, so there's friction when the bow is moved forward or backward. This friction can be increased by applying resin to the hairs. The hairs are kept tight by arching the wooden section of the bow.

The strings of the violin are stretched across the bridge from the tail to the pegboard. They are stationary at each end, so that standing waves are set up when they are bowed. The length and the amplitude of these waves depend on properties of the string such as its tension, its length, and the material it is made from (we discussed them in chapter 4). The first strings were made of catgut; they are now usually made of steel or various synthetic materials.

A casual look at a vibrating string seems to indicate that it is oscillating with a single loop, but a closer look shows that things are much more complicated than they appear. At any given time the string actually consists of two nearly straight segments; furthermore, these segments move around the curve at the frequency of the note that is being played, as shown in figure 79. To understand why this happens, we have to begin by pointing out that there are two types of friction known as static friction and sliding (or kinetic) friction. If you have a block sitting on the floor and try to push it, you find that it doesn't move until you have overcome the friction holding it in place. This is the static friction. Once the block begins to move, however, you notice that the friction decreases, and it doesn't take as much energy to keep it going as it did to start it moving. This is the sliding (or kinetic) friction. The difference between the two is important in the bowing of a violin.

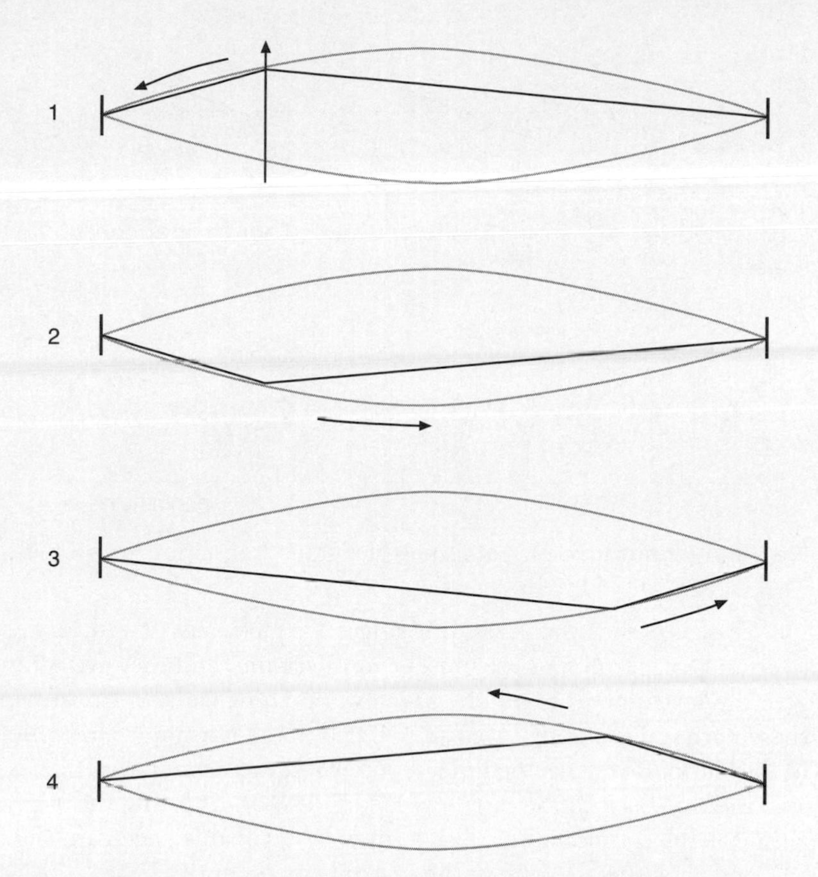

Fig. 79. Movement of the string as a violin is bowed.

When a player begins to bow, the bow grabs the string and pulls it to the side. At this point static friction is operating, but there is a restoring force in the string as a result of its displacement, and eventually the friction between the bow and string can no longer hold the string and it snaps back. As it moves back, kinetic friction, which is less than static friction, takes over, and the string slides easily under the bow. Soon, however, the bow picks the string up again as static friction takes over, and it begins to move at the same speed as the bow. As a result of this back-and-forth between static and kinetic friction, the motion of the string is periodic, but it is not simple harmonic motion. It is pulled at a constant speed (the speed of the bow), then stops suddenly, and snaps back, as depicted in figure 80.

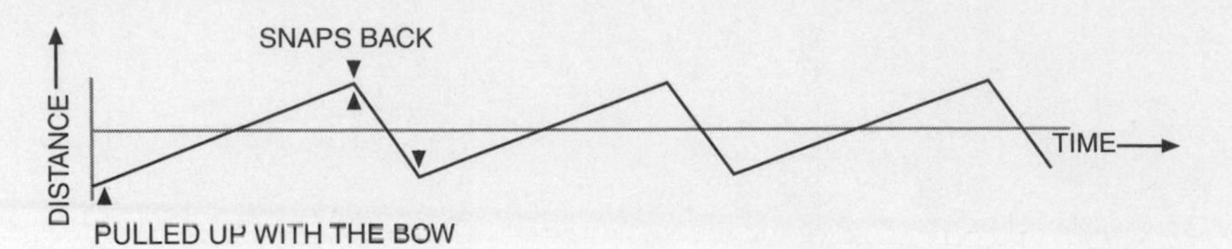

Fig. 80. Motion of a string as a bow is pulled across it at constant speed.

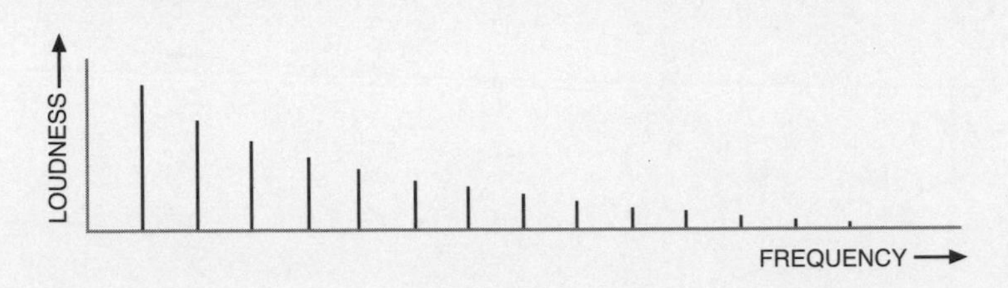

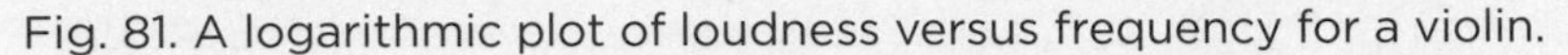

Fig. 81. A logarithmic plot of loudness versus frequency for a violin.

We saw earlier that even if a single frequency is bowed, many overtones or harmonics are present in the string, and they are what give the instrument its timbre. We also saw how we can determine these harmonics and that we can determine the harmonic spectrum of the violin. If we take the fundamental as 1, the second harmonic is 1/2, the third is 1/3, the fourth is 1/4, and so on. Plotting the logarithm of the harmonics against frequency gives this spectrum (fig. 81), and it tells us a lot about the instrument's sound.

Also of importance to any violinist is the *quality* of the sound produced. Quality depends on several factors, including

- the position where the string is bowed,
- the force of contact of the bow on the string,
- the speed of the bow,
- the width of the bow in contact with the strings,
- the spacing of the various resonant frequencies of the violin, and
- the physical construction of the violin—the materials out of which it is made.

The Stradivarius Secret

No one is sure where or when the first violin was made, but "fiddles" were known to exist in the thirteenth century. Andrea Amati of Cremona, Italy, is generally credited with making the first violin (as we know it today) in about 1550. He was the first of several generations of violin makers who by 1600 had made Cremona the unchallenged center of violin making in Europe. The Amatis are now credited with most of the major features of the violin—its shape, its overall structure, and its size. The grandson of Andrea, namely Nicolo, is now considered to have been the greatest craftsman of the family. For all its fame and influence, however, the Amati family was eventually overshadowed by a name that is still associated with great violins: Antonio Stradivari. Stradivari's violins are now the most prized musical instruments in the world.

Born in 1644 in Cremona, Stradivari had an uphill battle to distinguish himself in the trade of violin making. The Amatis dominated the scene, and the Amati violin was famous throughout Europe. This did not deter Stradivari; he worked for Nicolo Amati as an apprentice for a number of years and then in 1680 went into business for himself. He was determined to produce violins that were as good as or perhaps better than those of the Amatis. While his first violins were definitely inferior to those he made later, he continued developing his skills and experimenting with new models, and by 1700 he had hit his stride. During his life he produced about 1,200 violins. He also made 21 cellos (one of which played an important role in a Bond movie), 10 violas, and a few guitars. The years between 1700 and 1720 are now known as Stradivari's "golden years." About 600 of his violins have survived, and those from this era are invaluable, most of them worth millions of dollars. One sold in the United States in 2006 for $3.5 million, and others have sold for more.

What makes Stradivarius violins so valuable? As any violinist will tell you, they have a distinctive and beautiful tone. And strangely, even with all the technology we have today, no one has ever matched the "Stradivarius sound." What makes them so magical? What is different about them? There's no doubt that Stradivari was a master craftsman and took meticulous care to ensure that each of his violins was the best he could produce. He used spruce for the top; maple for the back, strip, and neck; and willow for most of the internal parts. He is also known to have treated his wood with several types of minerals. But there has to be more.

Many explanations have been put forward for the magnificent sound of his violins, but so far, none has proven to be correct. Stradivarius violins have been x-rayed, carefully measured, examined in intricate detail, and copied exactly, but the copies still don't sound like a Stradivarius. Is it the wood? The quality of the wood definitely affects the tone. The wood Stradivari used was particularly dense, and a dense wood is known to give a superior tone, but if this was all there was to it, why didn't other violins made in Cremona at this time measure up to a Stradivarius? Another idea was that it was the special varnish that Stradivari used. But the Stradivari violins are all several hundred years old, and most have been revarnished.

Stradivari's "secret" apparently died with him. His three sons did not continue in the trade; one died at 24, and another helped in his shop for a while but eventually left. It has been 250 years since Stradivari produced his last violin, and surprisingly, little has changed in the structure of the violin since that time.

Other Members of the Violin Family

The three other members of the violin family are the viola, the cello, and the double bass. Like the violin they all have four strings. They differ in physical size and in the fundamental frequency of their strings. Table 12 shows the fundamental frequencies and the lengths of the four members of the violin family.

The viola is tuned a fifth below the violin, and the cello a twelfth below. If you compare wavelengths on these three instruments with those on the violin, you would expect these instruments to be much larger than they actually are. From table 12 you see that C_3 on the viola replaces G_3 on the violin. If you compare the wavelength of the viola's C_3 string to that of the G_3 string on the violin, you will see that it is three times as long. Yet the viola is only 6 cm longer than the violin. Applying this same reasoning to the double bass, you would think that the double bass should be six times the size of the violin.

Table 12. Fundamental frequencies and lengths for instruments in the violin family

	Fundamental frequencies				Length (cm)
Violin	G_3	D_4	A_4	E_5	60
Viola	C_3	G_3	D_4	A_4	66
Cello	C_2	G_2	D_3	A_3	115
Double bass	E_1	A_1	D_2	G_2	200

This is not the case because compensations are made (such as using thicker strings), and the larger instruments are designed to be smaller for ease of playing.

The Virtuosos

We had an introduction to one of the violin's greatest virtuosos, Nicolo Paganini, at the beginning of this chapter. Let's look at his career in a little more detail and also at some of the other virtuosos. Paganini was born in Genoa, Italy, in 1782, and was quickly recognized as a prodigy. In many ways his story is similar to that of Mozart. His father recognized his talents and was eager to capitalize on them; as a result he forced his son to practice long hours every day (up to 10 hours and more). And indeed, his progress was dramatic. By the time he was 8 he was playing concerts around Genoa, and at 13 he was recognized as a talent without equal. Soon, he was playing throughout Italy to ever-increasing acclaim. At 17 he finally broke free from his father's overbearing dominance and went on his own, but by the time he was 19 he was gambling and drinking, and although he made a tremendous amount of money, he was soon broke. He was so broke, in fact, that he had to pawn his violin, and with a concert looming, he was forced to ask a wealthy French merchant if he could borrow a violin from him. The merchant loaned him a valuable Guarnerius (second to the Stradivarius in fame), then refused to take it back after the concert was over. Over the next few years Paganini used it extensively at his concerts. He was also the proud owner of a Stradivarius, which he won in a wager.

Paganini took some time off at the turn of the century but returned to the concert stage in 1805, and for the next several years performed throughout Italy. He suffered from Ehlers-Danlos syndrome, which, surprisingly, was helpful to him. The syndrome is marked by excessive flexibility of the joints, and this enabled him to perform astounding feats at the violin. His wrists were so loose he could move and twist them in all directions with ease, allowing him to do things other violinists could not do. His technique was, indeed, astounding.

A more recent virtuoso was Fritz Kreisler (1875–1962), who was born in Vienna, Austria, and studied at conservatories there and in Paris. He made his first tour of America in 1888–89. After this tour, and after being turned down for a position with the Vienna Philharmonic, he left music and studied medicine and painting for several years. In 1899, however, he returned to music and toured America for several years. He lived in Germany and France for a while, but spent the latter part of his life in the United States. Besides performing, he was also an ardent composer.

Jascha Heifetz (1901–87) also ranks as one of the great violinists of all time. A child prodigy, he was able to play violin concertos at the age of six. Born in Lithuania, he entered the conservatory at St. Petersburg in Russia at age nine; by the time he was 12 he was playing in Germany and Scandinavia. During his teens he toured much of the rest of Europe. Although he was primarily a soloist, he enjoyed playing chamber music with others and frequently played with pianist Artur Rubinstein. He came to the United States in 1917 and later taught at the University of Southern California.

No list of violin greats is complete without the name of Yehudi Menuhin (1916–99). He was born in the United States but spent most of his performing career in the United Kingdom. He performed for soldiers during World War II and played in Germany after the war. Although he was primarily a classical violinist, he also made jazz recordings and even made a few recordings of Eastern music with Ravi Shankar. In 1990 he was awarded the prestigious Glenn Gould Prize for his lifetime of contributions to violin music.

Another top violinist of the twentieth century was Isaac Stern. Born in Ukraine in 1920, he moved to the United States with his family when he was a year old. He studied music at the San Francisco Conservatory and made his debut when he was 16. He toured China in 1979 and made numerous recordings of violin concertos. He also dubbed the violin music in the movie *Fiddler on the Roof.*

Itzhak Perlman is one of the most famous among contemporary violinists. Born in Israel in 1945, he studied music in Tel Aviv, and later at Juilliard in the United States. He joined the Israel Philharmonic Orchestra in 1987 and toured with this group throughout Europe and in Russia, China, and India. He has appeared on U.S. television numerous times and has played at the White House. While his repertoire is primarily classical, he also plays jazz and has made a jazz album with the prominent jazz pianist Oscar Peterson. He has also been the soloist for a number of movie musical scores, including *Schindler's List* and *Memoirs of a Geisha.* He plays a Stradivarius violin.

Plucked Instruments: Banjos, Mandolins, Ukuleles, and Harps

The guitar is the best known and most popular of the plucked instruments, but we will leave it to the next section. Three other major plucked instruments are the banjo, the mandolin, and the ukulele. Their properties are listed below.

- Ukulele: 4 strings tuned to D_4, F_4, A_4, B_4
- Banjo: 5 strings tuned to C_3, G_3, D_4, A_4 (along with a short melody string)
- Mandolin: 8 strings (4 doubles) tuned to G_3, D_4, A_4, E_5

Like the violin, the banjo and the mandolin have a bridge, but the ukulele does not; its strings are attached directly to the resonant cavity. The mandolin and ukulele have a hollow body and a sound hole. The banjo, on the other hand, has a skin stretched over its upper end, like a drum; it amplifies the sound in the same way a resonant cavity does. The frets, fingerboard, and general overall construction of the instruments are generally the same, with the banjo being about 90 cm long, and the mandolin and ukulele being about 60 cm.

Another plucked instrument, quite different from these three, is the modern harp. Like the above instruments, it has a hollow soundbox (or more exactly, a soundbar), but it is much larger than the guitar family of instruments and, unlike them, has a vertical pillar, a curved neck, and 46 strings. Seven pedals that are connected to rods can be used to change the effective length, and therefore the pitch, of the strings. It is used extensively in modern symphony orchestras.

The Guitar

The guitar is similar to the violin in that it is basically several strings strung tightly over a sound box. There are two types of guitars, the acoustic guitar and the electric guitar.

Most of the basic characteristics of the violin also appear in the guitar. As in the case of the violin, the guitar is made up of top and bottom plates. The guitar needs a relatively large surface to push air back and forth, one that is relatively flexible so that it moves easily. Because of this, the top plate is usually made of a soft wood such as spruce, cedar, or pine. It is relatively thin (a few millimeters), so it is reinforced on the inside with a number of braces; their function is to keep the top plate flat. Most of the sound comes from the vibrations of the top plate. The bottom plate is usually made of mahogany or Brazilian rosewood, and it usually vibrates at lower frequencies than the top plate. The sides are also frequently made of mahogany or rosewood. Finally, the fretboard is usually made of ebony or maple.

But if the violin and the guitar are so similar, why do they sound so different? There are two reasons for this. First, they differ slightly in structure, and this difference has a significant effect on the sound. Second, the overtones in the two instruments are quite different.

Acoustic Guitars

There are two main types of acoustic guitars: the classical guitar with nylon strings and the folk guitar with steel strings. Both have six strings, which are tuned to the following notes and frequencies:

E_2	A_2	D_3	G_3	B_3	E_4
82 Hz	110 Hz	147 Hz	196 Hz	243 Hz	330 Hz.

These are, of course, the root (or fundamental) tones of each string. The strings also have many overtones, or harmonics, and these overtones give the guitar its particular sound.

If you look at a guitar, you will see a number of characteristics that make it distinctive:

- The strings are all the same length.
- The strings have different thicknesses.
- It has six strings instead of the violin's four.
- The frets along the fretboard are spaced differently.
- The sound hole is circular rather than f-shaped.

Let's consider the strings first. When a string is plucked, a standing wave is set up, and as we saw previously, it has a velocity v, a wavelength λ, and a frequency f, related by the formula

$$v = \lambda f.$$

Furthermore, the speed of the wave (v) depends on the tension of the string (T) and also on its mass per unit length (μ) according to the formula

$$v = \sqrt{(T/\mu)}.$$

We also saw earlier that the fundamental is twice the distance from the bridge to the nut or, when you press down the string, twice the distance between the bridge and the particular fret you are pressing down. Furthermore, all six strings have the same range of wavelengths. Looking at our formulas, therefore, we see that to change the frequency of a particular string we have to change the speed of the waves on it. This can be done in two different ways: by changing the tension T or by changing the mass/length μ (or by a combination of the two).

If we change the frequency by increasing the tension, however, the highest-frequency strings will be very tight and the lowest very loose, and this will make the instrument difficult to play. It would be much better if all strings had about the same tension. The only way we can do this is to change the mass/length ratio of the string (strings with higher mass densities have a lower frequency). And, indeed, this is what is done for guitar strings. On steel-string guitars the strings get thicker from high note strings to low note ones. In the case of classical guitars, however, the size change is complicated because of a change in density of the strings. The low-density nylon strings get thicker from E_4 to G_3, and the high-density wire-wound nylon strings get thicker from D_3 to E_3.

As we saw above, the fundamental notes on most guitars are E, A, D, G, B, and E. (There are guitars with 7, 8, 10, and 12 strings, but they are not common, so we will not address them here.) The other notes of the scale are obtained by pressing down on the string at the edge of a fret; this shortens the length of the string and increases the frequency of the fundamental (fig. 82). Along the neck of a guitar are a large number of frets that are set perpendicular to the strings and are usually made of metal. One of the first things you notice about them is that they are not equally spaced; the reason for the unequal spacing is related to the notes of the musical scale. When we press down behind a fret, creating a shorter length of string, we want

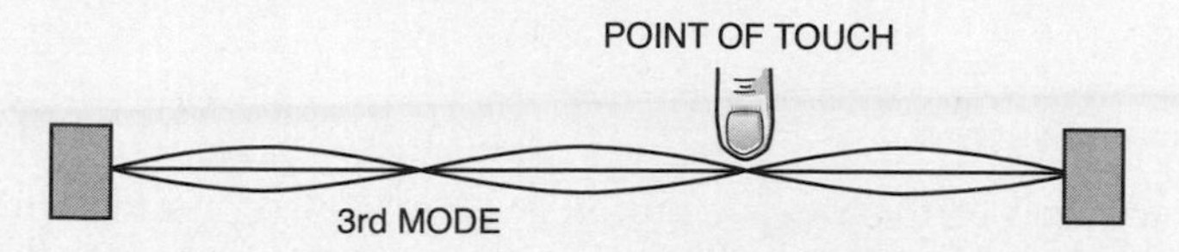

Fig. 82. Effect of pressing down on a guitar string.

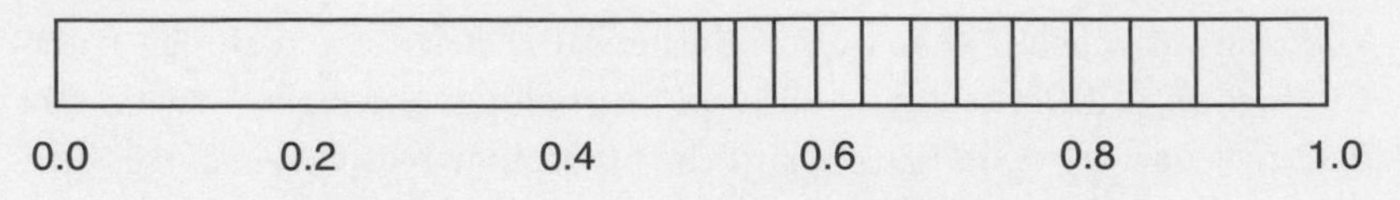

Fig. 83. Diagram showing the spacing of frets on a guitar.

the fundamentals to match (as closely as possible) the notes of the equal-tempered scale, which we discussed in chapter 5. We saw that it has 12 intervals making up the octave, and that the ratio of frequencies for half (or semi) steps is given by $r^{12} = 2$ which gives us $r = 1.0595$. From this we can easily determine the fret spacing that we need to give the proper scale. Assume that the open string length is l; the first fret must be placed at l/1.0595 from the bridge, the second at $l/(1.0595)^2$ from the bridge, and so on. In particular, the twelfth fret should be placed $l/(1.0595)^{12} = 0.5$, in other words, halfway along l. Figure 83 illustrates the spacing of the frets.

Now, let's consider how the guitar creates its music. As in the case of the violin, the strings themselves don't move enough air to be heard; their vibrations have to be transferred to something that moves more air. And this is, of course, the resonance box. Again, this resonance box doesn't amplify the sound in the usual sense. The energy imparted to the strings by plucking them is not increased; it is just converted to make it more efficient. Instead, there is a transfer of vibrational energy.

The energy of the strings is transferred via the bridge to the top plate, which, as we saw, is thin and flexible and therefore moves easily. The top plate, in turn, causes the air inside the body of the guitar to vibrate with a fundamental resonance called the Helmholtz resonance. Resonance patterns are, in fact, formed on the face of the top plate and are different for different guitars, depending on their construction and the materials used in them. The body, in effect, acts as a "resonant cavity," as in the case of the violin. It communicates with the outside air via a circular hole that is about 3¼ to 3½ inches across.

Electric Guitars

Now we turn to the electric guitar. It is, of course, quite similar to the acoustic guitar, except that it uses an amplifier rather than a resonance box to amplify the vibrations of the strings. The body of the electric guitar is a solid, shaped, flat board that has no acoustic function; it merely holds the strings and the electrical equipment needed for the guitar's function. Like the acoustic guitar, the electric guitar has six strings (in most models), which are plucked with a plectrum. Beneath the strings are two or more sets of electromagnetic "pickups" that convert the vibrations of the steel strings into electrical signals. These signals, in turn, are fed via a cable to amplifiers. The most common type of pickups are magnets that are tightly wound with copper wire. They work on the principle of the generator in that the vibrations of the strings create small voltages in the coils; these voltages are transmitted to the amplifiers.

Most guitarists employ devices to distort the sound. Two of the most common are the "fuzz box" and the "wah-wah." The fuzz box flattens the top of the signal waveforms and in the process adds additional harmonics. The overall effect is to make the sound fuzzy, as the name suggests. The wah-wah modulates the high harmonics up and down periodically, giving rise to a wah-wah sound.

Guitar Virtuosos

Finally, I have to say something about the guitar virtuosos. As you might expect, they are not generally associated with classical music. While there are and have been a number of classical guitarists, such as Andrés Segovia and, more recently, Julian Bream and John Williams, the guitar is primarily an instrument of popular music. It is, in fact, usually the lead instrument in the blues, rock and roll, and country music. I can't begin to cover all the well-known guitarists and will mention only a select few. In country music the best-known guitarist was Chet Atkins. Interestingly, he had little style of his own in his early years, but in 1939 he heard Merle Travis and was inspired by his finger-picking technique. Travis used his index finger to pick the melody and his thumb for bass notes. Atkins went further and developed a style of picking with his first three fingers, using his thumb for bass notes. He recorded many albums and also played backup for many well-known groups and singers. He later played a central role as a country music record producer, discovering many artists who later became famous.

In the areas of rock and roll and blues, no one is more admired than Jimi Hendrix. Many claim he was the best electric guitarist who ever lived. Certainly, he had a lot of talent and introduced a lot of innovations to guitar playing. He used effects such as the wah-wah and distortion extensively but tastefully.

Next on my list is Eric Clapton. He dabbled in several different kinds of music but is generally associated with rock and the blues. Early on, he was associated with the band The Yardbirds, but later joined the trio Cream. He has been a powerful force in rock music for almost four decades and is considered to be one of the great guitarists.

Finally, I must mention the "King of the Blues," B. B. King. King was best known for his single-note solos, played on his hollow-body Gibson guitar. His tone has been described as "velvety," and he was well known for his "trilly" vibrato. He nicknamed his guitar "Lucille." Later, the Gibson company (with his permission) marketed a guitar designed after King's guitar and called "Lucille."

10

The Brass Instruments

Trumpet and Trombone

It is no exaggeration to say that one of the greatest trumpeters of all time, Louis Armstrong (fig. 84), had a rocky start in life. Shortly after he was born in 1901 in New Orleans, his father left, and his mother sent him and his sister to live with their grandmother. With money always in short supply he had to go to work at an early age, cleaning fishing boats, selling newspapers, and selling coal from a cart. At the age of 12 he was sent to reform school for firing a gun in the air during a New Year's Eve celebration. Interestingly, it was here that he first came in contact with a cornet, and he soon fell in love with it. He continued playing it throughout his school years; although he had a few lessons, most of what he learned was from local musicians.

Despite numerous handicaps in his early life he persevered and went on to become one of the most respected and beloved jazz musicians of all time. He made hits of such songs as "When the Saints Come Marching In," "St. Louis Blues," "Stardust," "Ain't Misbehavin'," and "What a Wonderful World." In 1964, his recording of "Hello Dolly" went straight to the top of the charts, and at 63 he became the oldest person to ever have a number one hit. Along the way he appeared in numerous movies and worked with many well-known

Fig. 84. Louis "Satchmo" Armstrong.

musicians, including Bing Crosby, Ella Fitzgerald, Earl Hines, and Jimmie Rodgers.

I never heard Louie Armstrong in person, although I saw many of the movies he was in. Nevertheless, I did have the pleasure of hearing another great trumpeter, Al Hirt, and was enchanted by his renditions of "The Carnival of Venice" and "The Flight of the Bumblebee." Hirt was also from New Orleans and began playing the trumpet at the age of six. I think the trumpet became one of my favorite instruments after I heard Bert Kaempfert's "Wonderland by Night." It, along with "Cherry Pink and Apple Blossom White," are still two of my favorite trumpet pieces.

The trumpet was one of the earliest musical instruments. According to ancient records the Chinese had a form of the instrument as early as 2000 BC. And, of course, the trumpet is referred to several times in the Bible. The walls of Jericho fell down when trumpets were sounded. Trumpets were also used in Greece in connection with the Olympic Games in 400 BC.

The earliest "trumpets" were horns and were likely made of animal horns. In Switzerland, however, the first horns were made from trees, and were called *alphorns;* they were particularly long, sometimes as long as five meters (see fig. 85) and produced only very low notes. They were used mainly for calling the cows home in the evening so would hardly qualify as a musical instrument. It wasn't until the development of metals, and metal working, that horns were made of brass. According to early records, the Egyptians were one of the first to make brass horns, or what we might call bugles, in about 1500 BC. These horns were used mainly by the military.

The early horns, however, were quite different from the trumpets of today. The thing that revolutionized the instrument was the invention of valves. The first valve trumpet was introduced in Berlin in 1814 by Heinrich Stölzel. The earlier trumpets had been restricted to a few notes, but the new valves made it possible for musicians to play all the notes of the chromatic scale. The first valve trumpet was called a flügelhorn.

Fig. 85. An early Swiss alphorn.

Vibrating Air Columns

We talked about vibrating air columns within tubes in chapter 4, but to refresh your memory I will review this topic briefly. As we saw, there are two cases: tubes with both ends open, and tubes with one end closed and one open. One of the key things we noted earlier was that the frequency of vibration within a tube depends on its length. The longer the tube, the lower the frequency; for the most part this frequency is independent of the diameter of the tube, as long as it isn't too large.

In the case of the tube with both ends open, the ends are, of course, at atmospheric pressure, and they have to stay at this pressure when a wave is set up within the tube. This means that significant changes cannot occur at the ends. Changes of pressure do occur inside the tube, however, and these changes are similar to those that occur in a vibrating string that is fixed at both ends. We can, in fact, represent them on a diagram in the same way (fig. 86). As you see in the diagram, large pressure changes occur away from the open ends, with the largest being at the center of the tube. However, this is a plot of pressure versus length along the tube, and it is showing only the extremes of pressure. Most of the time the pressure is between these two extremes. If you compare this with the case of a vibrating string, we know that the string is only at its extreme position a small amount of the time; most of the time it is fluctuating back and forth between these extremes. In the same way, the pressure in a tube is also fluctuating back and forth most of the time.

Brass instruments such as the trumpet do not have two open ends, of course. The player's lips are over one end, and this gives a tube

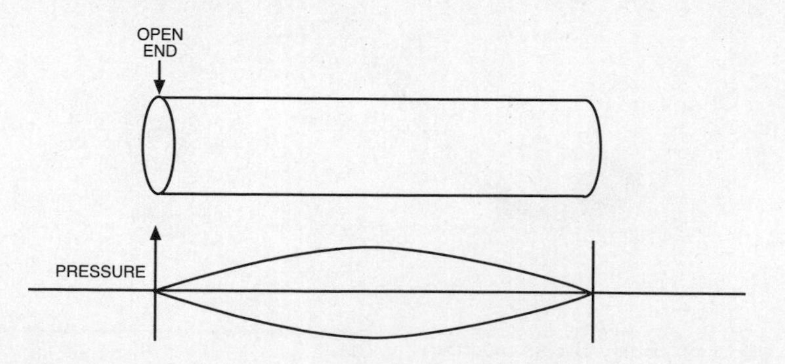

Fig. 86. Wave in a tube with two open ends.

with one end closed and one open. So let's turn to this case. As the diagram in figure 87 shows, there is a significant change in this case. The pressure at the closed end is no longer restricted to air pressure; in fact, the largest fluctuations occur here. From the diagram we see that the wavelength in this case is twice as long as it was in the case of two open ends (fig. 86). The frequency is therefore lower. In fact, the closed end causes the frequency to drop by an octave (2:1) compared with the open-ended case. The major change occurs at the closed end—in other words, at the player's lips. His lips are vibrating rapidly, and this is, of course, what is creating the vibrations within the tube. (We will look at this in more detail later.)

In figure 87 we see only a quarter of a wavelength, but we know that many higher waveforms can occur, namely, overtones or higher harmonics. In the case of a tube with two open ends we have the same situation. The lowest, or fundamental, wave was half a wavelength, but we can also have two, three, four, and more half wavelengths along the tube. In particular, we see that the first overtone is twice the fundamental, and the second is three times the fundamental, and so on (fig. 88). This gives a harmonic series. If we compare this open-ended tube with the closed end tube, we see that in the tube with one closed end the wavelength of the first overtone is longer for the same tube; it is, in fact, a half a wavelength longer but the wavelength has changed (fig. 89). In short, we have two loops for the open case and one and a half loops for the closed case. The length is therefore two-thirds the length of the wave in the closed tube wave. The frequency of the wave in the open-ended tube is therefore three times that of the fundamental, and this means there are no second harmonics.

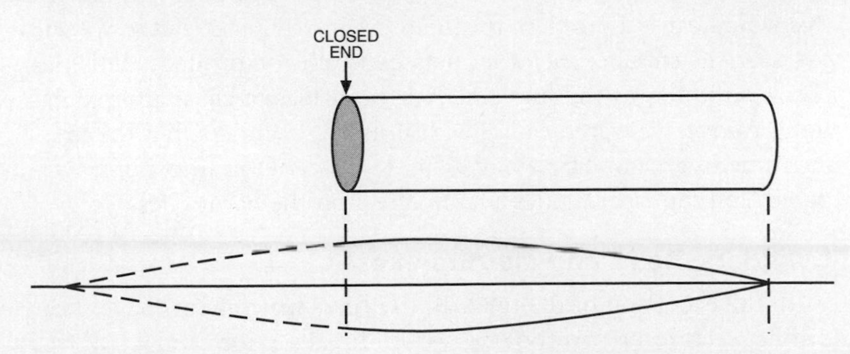

Fig. 87. Half a wavelength in a tube with one end closed.

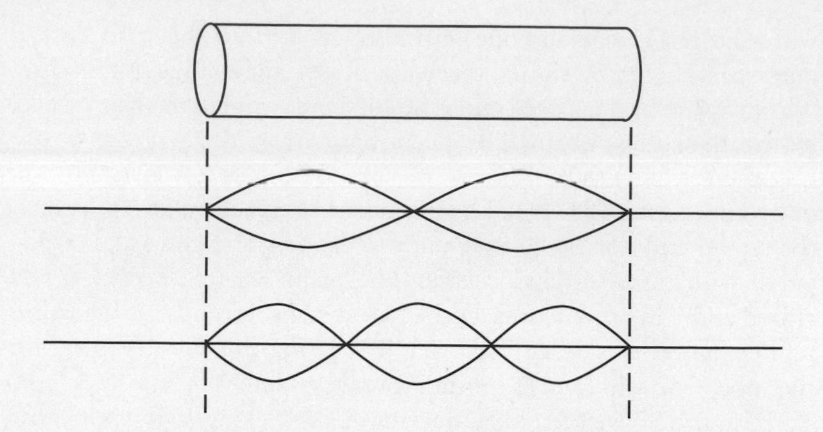

Fig. 88. First and second overtones for a tube with two open ends.

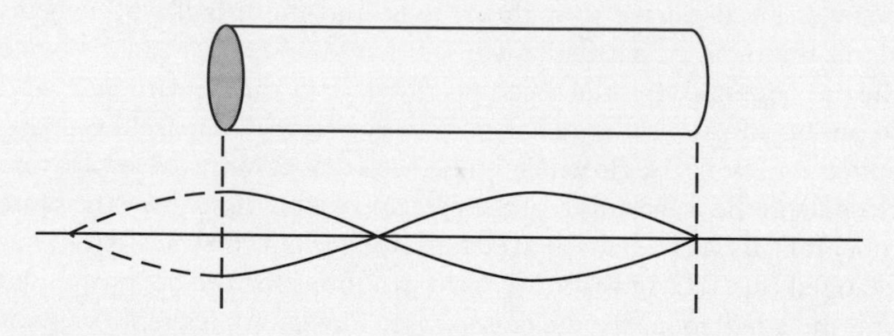

Fig. 89. First overtone for a tube with one end closed.

We can, in fact, continue in the same way and show that the closed-end tube has only odd-numbered harmonics. You might think that this would cause a problem for the brass instruments, but as we will see later, the trumpet and other brasses are able to produce all the integral multiples of the fundamental (or at least a close approximation). Above, we were dealing with uniform cylinders, but the brass instruments are not uniform in diameter; part of the air column is tapered, and the tapering has a large effect on the harmonics.

Cylinders, Bells, and Mouthpieces

With the exception of the tuba, all the brass instruments have three distinct sections: a mouthpiece, a cylindrical section, and a bell (see fig. 90). Each of these sections influences the acoustics of the instrument. To illustrate, let's consider the trumpet. For now we'll ignore the fact that it has three valves and consider it to be a tube (as illustrated in the figure). In the case of the B-flat trumpet (the most common of the trumpets) the fundamental is at a frequency of about 115 Hz. It is generally too low to be played but does have an important role in relation to the overtones. The overtones are integral multiples of this fundamental, as shown below.

Overtone	*Frequency*
1	2 × 115 = 231 Hz
2	3 × 115 = 346 Hz
3	4 × 115 = 455 Hz
4	5 × 115 = 570 Hz
5	6 × 115 = 685 Hz

The overall length of the trumpet is approximately 140 cm, so let's consider a cylindrical tube 140 cm long that is closed at one end. As we saw, the resonant frequencies are odd multiples of the fundamental, which in this case can easily be shown to be 62 Hz (using the formula $v = \lambda f$). We therefore get the harmonic series 62, 186, 310, 434, 558, and 682 Hz, which is not even close to the frequencies of the trumpet that are listed above. The reason, as I mentioned earlier, is that although part of the trumpet is cylindrical, it has a tapered mouthpiece and a tapered bell, and they must be taken into consideration.

Consider the bell first. It has a large influence on the sound, and its shape and taper are critical. If, for example, it had a smooth, gradual taper throughout its length, it would merely be a megaphone; the traveling waves inside it would move directly out of the bell; and there would be no standing waves (which are needed for it to be a musical instrument). On the other hand, if the horn had no flare, it

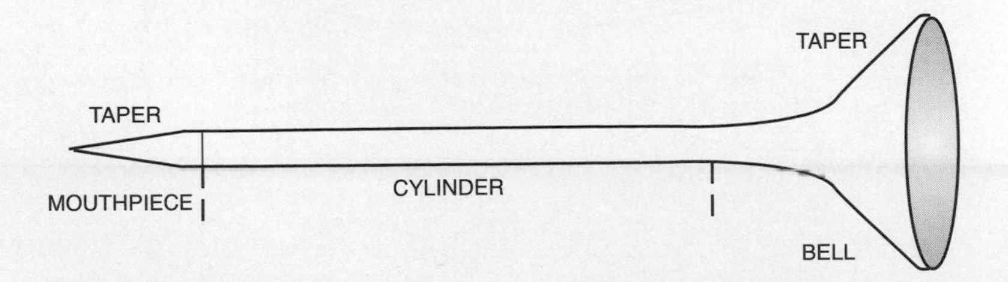

Fig. 90. The three main sections of a brass instrument.

would reflect all, or at least most, of the waves, and none would leave at the open end. We would therefore be unable to hear the sound.

What we want is a bell that creates standing waves within its length, but allows some of the sound energy to travel out at the open end. In the trumpet, most of the vibrational energy is in the form of standing waves, but some "leaks" from the bell to the open air. For this to happen, the bell has to be tapered in a particular way. Near the end of the bell there is a node, and its position changes as the wavelength of the wave inside the trumpet changes. The waves are reflected from this node. Analysis shows that the node occurs at the point where the rate of flare is rapid compared to the wavelength of the wave. This means that at low frequencies the node is deep within the bell, but at higher frequencies it is close to the end (fig. 91).

In the same way, the mouthpiece also affects the frequencies. At low frequencies it adds to the length of the tube and lowers all frequencies. At higher frequencies, it has its own resonant frequencies, and these resonant frequencies add to the apparent length of the tube at high frequencies. In fact, the higher the frequency, the longer the tube (plus mouthpiece) appears.

So, both the bell and the mouthpiece affect the harmonics, and the effect is so great that instead of just odd multiples of the harmonics we get approximate integral multiples.

The resonant series for the trumpet is shown in figure 92. It allows

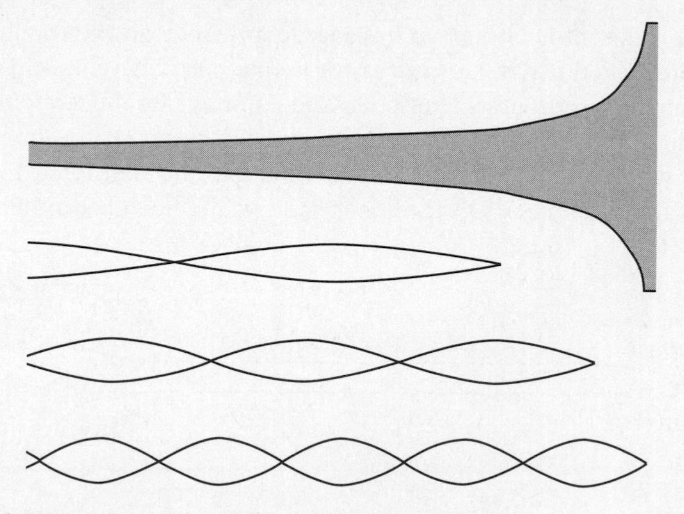

Fig. 91. Waves in bell end of a brass instrument.

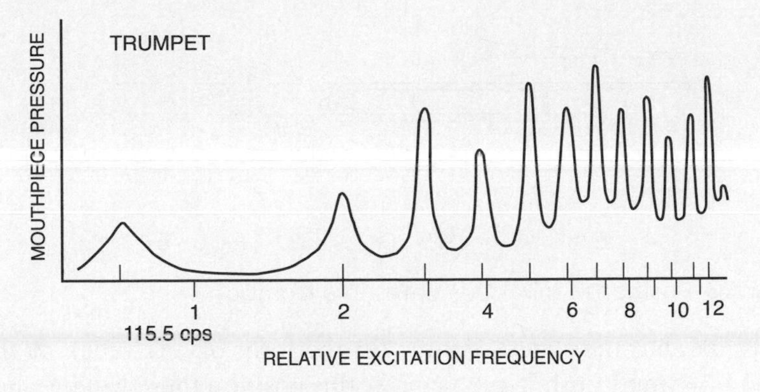

Fig. 92. Resonant series for a trumpet.

the trumpeter to play C_4, G_4, C_5, E_5, G_5, and a few higher notes. These are the only notes a bugle can play, and because of this all military pieces (think of "Taps") consist only of these few notes. For musicians playing in an ensemble or playing a solo, however, it is important to be able to fill in the notes between—in other words, to be able to play a chromatic scale. This also applies to the trombone and other brass instruments. To see how this is achieved, let's begin with the trombone.

Slides and the Trombone

As we saw, the harmonic frequencies in any brass instrument are determined by the length of the air column within it. Change the length of this column and you change the frequency, and this is what we use to bridge the gap between the frequencies f_2 and f_3 above (f_1 is below the usual range). In the trombone there are six half tones between these two resonances; thus, we need six different additional lengths to bridge it. Our problem is therefore to determine what the additional lengths should be. In the case of the trombone we add additional lengths using a slide (actually, it's a U-tube, or U-slide, since it doubles back on itself).

Let's begin by assuming the slide is brought to its shortest position. At this position the trombone has resonant frequencies of approximately 116, 174, 233, and 292 Hz. To lower the frequency by a half tone we have to decrease it by approximately 6%. Assume the length of the tube at its shortest position is 270 cm; this is typical of a trombone. To lower the frequency by a half tone we must add (270×0.06)

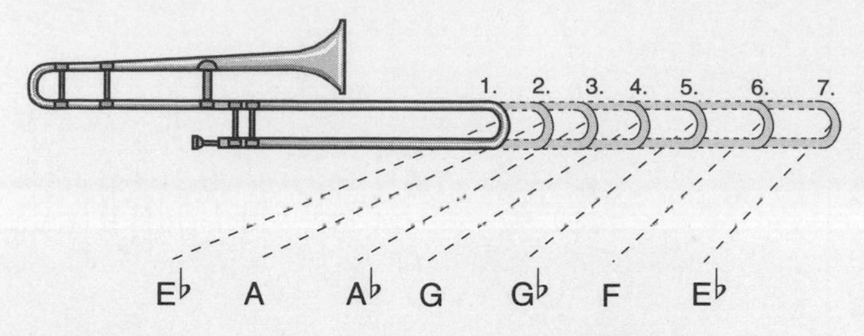

Fig. 93. Positions for various notes in a trombone.

= 16 cm, but the slide is double (in the form of a U-tube), so this means we must push it out 8 cm. At this position there will be a new family of resonances: 110, 165, 220, and 277 Hz.

The new length is now (270 +16) = 286 cm. To lower the pitch another half tone we must again add 6%, which is (286 × 0.06) = 17 cm, and this means extending the slide out another 8.5 cm. Again, we get another set of resonances. Continuing in this way we can cover all the notes of the chromatic scale between the resonances f_2 and f_3. All the notes of the scale can therefore be obtained by moving the slide to various positions, as shown in figure 93.

Valves and the Trumpet

In the case of the trumpet, valves rather than slides are used to extend the length of the air chamber, and they are a little more complicated. A single valve could, of course, be used to lengthen the chamber, but it was shown early on that three valves worked much better. As in the case of the trombone, to get to the notes between the fundamentals f_2 and f_3 we have to lengthen the air chamber. The valves are therefore connected to small U-tubes that increase the overall length of the air column. When the valve is in its upper position, air goes directly through the trumpet, but when it is depressed the air is redirected into the U-tube before it continues to the rest of the air chamber, as is shown in figure 94.

We'll assume the overall length of the air column inside the trumpet is approximately 140 cm when all valves are open (at their upper position). As in the case of the trombone, we can easily calculate what additional lengths are needed. Again, we need an increase of 6% for the first note, and (140 × 0.06) gives us 8.4 cm. This lowers the frequency by a half tone, and the valve for doing this is the middle valve in the diagram (fig. 95).

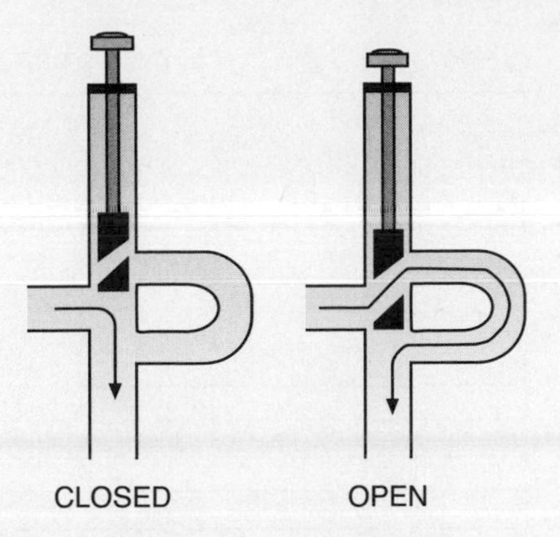

Fig. 94. Cross section of valves in a trumpet.

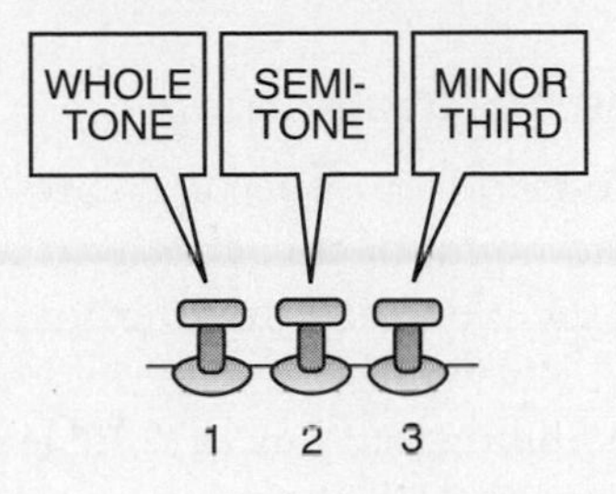

Fig. 95. The three valves of a trumpet. The first valve lowers the frequency by a whole tone, the second by a half tone.

For a whole tone we need an additional 12% increase, and (140 × 0.12) gives 16.8 cm, so we would need an additional 16.8 cm to lower it a whole tone. There is, unfortunately, a problem with three valves. The extra length that each one adds should be independent of the position of the other valves, and it is not. For example, the extra length that valve 1 adds is slightly different, depending on whether valve 2 or valve 3 is depressed. Also, we know that valve 3 lowers the tone by three half tones, but depressing both valves 1 and 2 also lowers it by three half tones, and unfortunately, they do not agree exactly. Because of this, slight adjustments or compensations have to be made in the "crooks" of the U-tubes, as shown in the schematic of figure 96. Small adjustable slides are therefore incorporated into the tubes to help musicians tune their instrument.

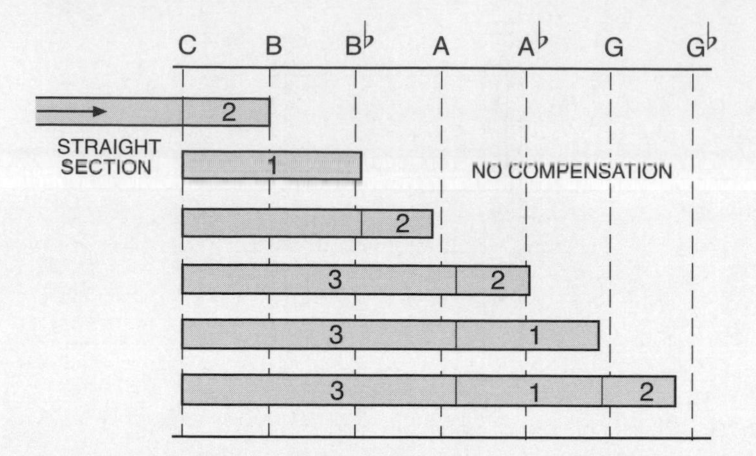

Fig. 96. Extra lengths in a trumpet give different notes.

A Sampling of Famous Trumpeters

We had a brief introduction to one of greatest trumpeters, Louis Armstrong, at the beginning of this chapter. He began his musical career by playing on the New Orleans riverboats that toured up and down the Mississippi River. But he soon discovered that New Orleans had limited possibilities, and in 1922 he left to join Joe "King" Oliver's band in Chicago, which was now considered to be the major center for jazz in America. He made several recordings there, but soon New York beckoned, and in 1924 he left to play in Fletcher Henderson's orchestra, the top African American band of the day. Surprisingly, he stayed in New York only a year before returning to Chicago. He was now making a large number of recordings and was becoming increasingly well known. By 1929 Armstrong was, in fact, becoming a star in the music world, and he soon formed his own band. In 1930 he moved to Los Angeles, where his career really took off. Over the next few years he had a heavy schedule, playing over 300 concerts a year; in addition, he appeared in many movies (30 in all).

Early in his career he played the cornet but soon switched to the trumpet. His powerful style eventually damaged his lips, and he began to sing more and more at his concerts. His raspy voice was quite different from that of most of the "crooners" of the time, but he eventually became as popular for his singing as he had been for the trumpet. He introduced a style called "scat singing," in which nonsense words and syllables are used (for instance, "bippity-doo-wop-bee-bop"). The style had actually been used in a limited form earlier by artists such as Al Jolson. Bing Crosby was one of many singers who were influenced by Armstrong; he also used scat frequently.

Also from New Orleans was Al Hirt. He began playing the trumpet at a young age and was soon playing in the New Orleans Junior Police Band. By the age of 16 he was playing professionally, but he took some time off in 1940 to study at the Cincinnati Conservatory of Music. He was a bugler in the army during World War II. After the war he played in several well-known bands, including those of Tommy Dorsey, Jimmy Dorsey, and Benny Goodman. He finally returned to New Orleans and formed his own band. During the 1950s and 1960s he had many albums on the pop charts. In all he recorded 22 albums, and his single "Java" went to the top of the charts; he later received a Grammy award for it.

One of the most popular trumpeters of the World War II years was Harry James. As was true for Louis Armstrong, James's early years were quite unusual. Both of his parents were circus performers—his father a bandleader and his mother a trapeze artist. His father played the trumpet and started his son on the instrument almost as soon as he could hold it, and young Harry was soon playing in the circus band. His progress was so rapid, in fact, that by the time he was in high school he was the star soloist at many concerts.

In 1936 James joined Benny Goodman's orchestra, where he became very popular with the crowds. In 1938 he decided to form his own band; a year later he hired an almost unknown singer named Frank Sinatra. He later also hired Dick Haymes as a singer. In the early 1940s he went to Hollywood, where he met and married the actress Betty Grable.

There were, of course, many other great trumpeters. Dizzy Gillespie was well known for his phenomenal technique. Perhaps one of the greatest was Bix Beiderbecke, who trained Bobby Hackett, Red Nichols, and Jimmy McPartland. One of Beiderbecke's most celebrated pieces was "Singing the Blues." The best-known living trumpeter is Wynton Marsalis of New Orleans.

The trumpet has also played a relatively important role in classical music. Some of the most famous classical pieces written for it are Franz Haydn's Concerto for Trumpet and Orchestra in E-flat major,

Paul Hindemith's Sonata for Trumpet and Piano, and Alexander Arutiunian's Concerto in A-flat for Trumpet and Orchestra.

The Other Brass Instruments

There are, of course, several other brass instruments, the major ones being the cornet, the French horn, the tuba, and the sousaphone. The trumpet is usually considered to be the "soprano" of the brass family. The cornet is similar to the trumpet, with the same range of frequencies, but is not as brilliant or "brassy" because it has fewer overtones. Any musician who plays one can play the other. Some musicians prefer the cornet, others the trumpet. For example, Bix Beiderbecke, Bobby Hackett, and Red Nichols preferred the cornet, while Louis Armstrong, Al Hirt, and Harry James preferred the trumpet.

The French horn is easily recognizable by its large circular coils. It has a frequency range from B_1 to F_5 and an air column length of 325 cm. It uses a different type of valve from that used for the trumpet; called a rotating valve, it is controlled by a small knob. The mouthpiece of the French horn is also different from that used by the other brass instruments: rather than being cup-shaped, it has a gradual and smooth transition from the lip rim to the back bore.

The range of the tuba varies, depending on the type; in the E-flat tuba it is from E_2 to $B\flat_4$. It has an air column length of 536 cm and is considered to be the lowest-voiced member of the brass family. Finally, the sousaphone is a portable version of the tuba, the large instrument you usually see in marching bands. It has a range from C_1 to A_3.

11

The Woodwinds

Clarinet and Saxophone

When I was about ten, I would occasionally hear music coming from the shed of one of my neighbors. It sounded like he was playing a musical instrument, but I had no idea what kind of an instrument it was. Every time I heard him playing, I would lean over the fence and listen. Finally one day he spotted me through a window and invited me to come over. I looked at the instrument as I entered the shed but still had no idea what it was. It looked like a large curled pipe with numerous gadgets on the side. He told me it was a saxophone and played several pieces for me; he also mentioned that he played in a small band. Needless to say, the experience left quite an impression on me.

The saxophone, along with the clarinet, are two of the most popular woodwinds and have both been used extensively in jazz bands, and even by rock and roll groups. The clarinet first came to my attention when I saw the movie *The Benny Goodman Story.* In a scene that I particularly remember, Goodman was playing the clarinet as he conducted his orchestra. Behind him the dance floor filled with dancing couples, but suddenly the dancing stopped as everyone began watching and listening to Goodman play. When he turned around, he was surprised.

Something close to this did indeed happen in 1935 at the Palomar Ballroom in Los Angeles. For the first night or so of their engagement Goodman and his orchestra played their usual stock arrangements, but times and tastes were changing, and the crowd's reaction to these selections was relatively cool. Famed drummer Gene Krupa, who was in the band, finally took Goodman aside and said, "Benny, we're going to die, so let's die playing our own thing." By this he meant a new form of music that would later be called "swing." It was much more upbeat and jazzy than what they were playing. Benny agreed, and they switched to the new swing arrangements. Soon the dancers began to cheer, and some even broke out in applause. Finally they stopped dancing and gathered around the bandstand listening to the music.

One of the outgrowths of the new, more upbeat music was a new dance style called "jitterbugging." It caught on fast, and interest in it soon spread. Within days, newspapers across the country were carrying stories about the new type of music and the new dance. And so began the "Swing Era."

Getting a Tone

The woodwinds include several instruments besides the clarinet and saxophone; the major ones are the flute, the oboe, the bassoon, and the recorder. "Woodwinds" might seem a misnomer, since several, including the flute and saxophone, are made of metal, but most were originally made of wood. Anyway, let's look at how we get a tone from them. Two of the easiest ways to produce a tone are by blowing across the mouth of a bottle (or something similar to a bottle) and by blowing through reeds. One of the simplest reeds is a piece of cardboard; if you place it up to your mouth and blow, you get a vibrating sound. It might not sound very musical, but as we will see, blowing through a reed is an important way of making a tone.

If you listen carefully to the sound produced in each of the above cases, you can hear that it is not a pure tone. It sounds like a jumble of many different tones, and indeed it is a chaotic array of waves of many wavelengths. It might seem, therefore, that these sound-producing techniques would be of little value in setting up standing waves, which are needed for a tone to have any musical value. But if you look at what happens to these waves in either a double, open-ended tube, or a tube with one end closed and one open, you soon realize that this is not the case. If the tube has one end closed, for example, the array of waves will move toward the closed end and will be reflected from it. They will, in fact, continue moving back and forth in the tube, and as they do, they will begin to change. As we saw earlier, any tube has a fundamental wavelength, and some of the waves in the array will likely be close to the fundamental. As the waves continue to move back and forth, the ones close to the fundamental will build up and the others will decay. Eventually, we will get a standing wave equal to the fundamental wavelength.

The two types of tones—those created by an edge and those created by a reed—are referred to as edge tones and reed tones, respectively. Reed tones can be generated by single or double reeds. Standing waves can, of course, be set up by either method. The flute, for example, uses an edge tone, the clarinet and the saxophone use a single reed, and the oboe and the bassoon use a double reed. Reed instruments are generally louder than edge instruments, other things being equal.

Using either source we can set up a standing wave in the bore of any of the these instruments, and as is the case with the brass instruments, the standing wave will have a fundamental and several overtones. We therefore have the same problem we had with the brass instruments: the overtones give a few notes of the scale, but for the entire scale we need more.

Filling in the Gaps: Getting a Scale

The steps we have to take to get a scale involve the shape of the bore of the instrument. A flute has a uniform bore and two open ends; the bore of the clarinet is also uniform, and as we saw earlier, it has an open and a closed end. Even a casual look at a saxophone, however, shows that its bore is not uniform; the size of the bore increases significantly toward the horn end. In short, it is tapered with one end closed and one open. The bores of the oboe and bassoon are also tapered.

Assuming now that we have a uniform tube, we can set up a series of standing waves in it corresponding to its fundamental and overtones. They will give us a few notes of the diatonic scale, but how do we get the others? If we have a tube that is open on both ends, we know that a half a wavelength will fit into it and is therefore the fundamental for the tube. What we need now is to decrease the "effective" length of the tube to increase the pitch of the waves inside it. How do we do this? It's natural to try drilling a small hole along its length. If this hole is exceedingly small, it has little effect on the tube, but if it is moderately large, it changes the wavelength of the wave in the tube and therefore changes the pitch. In effect, a hole makes the tube equivalent to a shorter tube (without a hole), which is exactly what we need.

Now, assume that we have several tubes of a given length and that we drill holes about half way along them, gradually increasing the size of the holes (as in fig. 97). We see that the "equivalent" tube (a tube with no holes that would have the same wavelength) gets shorter and shorter, until finally when the hole is roughly the diameter of the tube, the equivalent tube is equal in length to the distance from its base to the hole.

This means that by placing a hole along the tube we can get extra notes, with the pitch depending on the size of the hole. We can also change the effective length of the tube by placing more than one hole along it. For a tube open at both ends with a uniform diameter we can, in fact, get a diatonic scale with six holes along its side. The flute is an example of this case; if we start with all holes covered and uncover them one at a time beginning at the lower end we gradually increase the pitch, and with the appropriate holes we can get all the notes of the diatonic scale (fig. 98). We can, for example, go from C up to B, but to get C′ we have to cover all the holes and blow harder to set up the second harmonic. Blowing harder to get higher harmonics, or

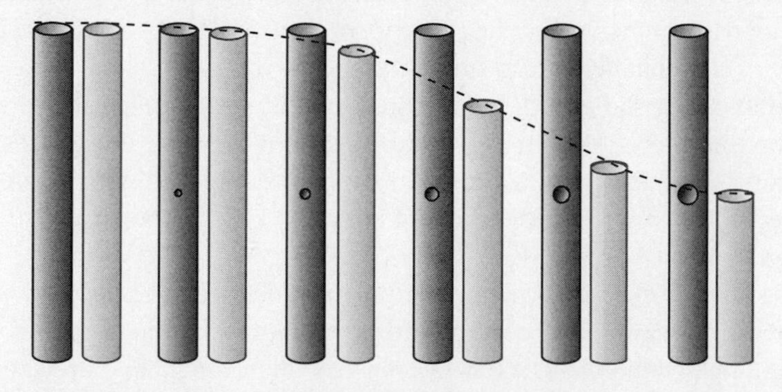

Fig. 97. The effect on wavelength of increasing the hole size in a tube. The tube on the right in each pair of tubes is the "equivalent" tube length without a hole.

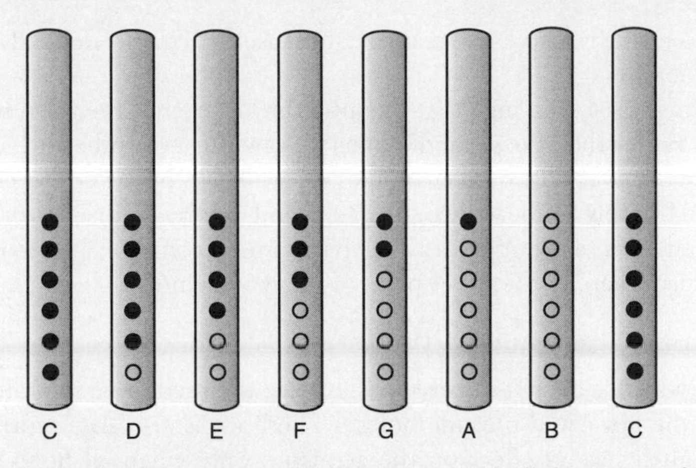

Fig. 98. By covering various holes we can get all the tones of the scale in an open-ended tube.

overtones, is referred to as *overblowing*. Once you get C′, you can continue uncovering the holes in the same way to get higher notes.

The overblowing technique works well with the flute, which is open at both ends, but it does not work with the clarinet, which has an open end and a closed end. In the case of the clarinet it's almost impossible to get to the first overtone by overblowing. To get around this, the clarinetist uses what is called the speaker key—a small hole on the bottom side of the instrument, usually about 15 cm from the mouthpiece. When the speaker key is open, it excites the third harmonic (fig. 99). Actually, it doesn't do this without a little help; you

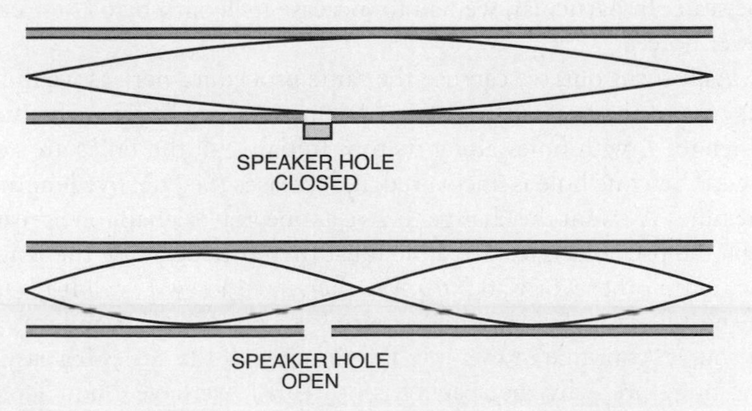

Fig. 99. The speaker key (hole) in the bottom of the clarinet.

have to learn how to use it properly because the change depends on how you blow.

It might seem strange that opening the key excites the third harmonic rather than the second, but as we saw earlier, a tube with one closed end gives only odd harmonics. As we'll see, however, the clarinet isn't simply a tube with an open end and a closed end; it has other parts, and as a result, it gives all the harmonics except for the second. Let's turn, then, to the other parts of the woodwinds.

The Total Instrument

All woodwinds consist of three basic parts: the reed or edge mouthpiece, the bore, and the side holes. (The horn is also an important part, but it has an effect on the acoustics only when all holes are closed.) When the instrument is played, the initial vibrations are created by the steady stream of air blown by the player. The vibrations themselves are set up by the mouthpiece, which may be either a reed or an edge. The frequency or pitch at which the air vibrates is determined mainly by the length of the bore, or more exactly, its effective length, which is determined by the number of covered or uncovered holes along the bore.

If you look at the holes of a clarinet or saxophone, you see that they are not separated by equal distances and are not the same size. Furthermore, all the woodwinds have more than six holes. Let's consider their varying distance first. With woodwinds, we have the same problem that we had with the trombone (or trumpet). In that case we had to add various lengths to the air chamber to get all the notes of the scale. In particular, we had to increase its length by 6% for each lower note.

As it turns out, we can use the same procedure here as we did in the case of the brass instruments. Assume that we begin with a tube of length L with holes along its top. Initially, all the holes are covered; when one hole is uncovered, it decreases the effective length of the tube. We want the change to give a tone that is a half tone greater than the previous one. The hole must therefore decrease the length by 6% (in other words, $0.06L$). We then have a new L; call it L'. For the next half tone we have to decrease L' by 6% ($0.06\,L'$), and we can, of course, continue in this way for all notes of the scale. It's easy to see, therefore, why the spacing is not equal: we have a new length each time. We saw the same thing earlier in the case of the spacing of the frets on the guitar.

The second thing we notice about the holes in clarinets and saxophones is that their size changes. This doesn't apply to all the woodwinds; the holes in the flute are all the same size. The most obvious case where they change is where the bore of the instrument changes in size. The reason for this is that the ratio between the hole size and the bore size must be kept constant if they are to be equivalent to a uniform set of holes down a uniform bore. If the bore changes, therefore, the hole size also has to change.

In the case of the clarinet, however, the bore is uniform; nevertheless, the holes still change in size. The reason for this is that the variable spacing has an effect on their size: the closed holes add to the effective cross-sectional area, and compensation must be made.

The Woodwind Instruments

The Edge Instruments: Flute, Piccolo, and Recorder

The flute uses an edge to produce its tone. The player blows across the opening, and the stream of air he or she produces sets up turbulence as it crosses the edge of the opening, with part of the air going into the opening and part escaping. The tone produced depends on how the air is blown and, particularly, on how the lips and tongue are positioned relative to the opening.

The flute is made in two keys, C and G, and is about 66 cm long. The concert flute, which is quite common in large orchestras, is usually made of silver with some gold. Student flutes, on the other hand, are made of an alloy of silver and nickel. The C flute has a range of C_4 to C_7, and the G flute has a range from G_3 to G_6.

An instrument closely associated with the flute is the piccolo, which is about half as long and is pitched an octave higher. The C piccolo has a range of D_5 to A_7.

Another edge instrument that is similar to the flute is the recorder. It is the instrument that is frequently used in elementary schools to introduce music to children. It has a line of holes along the top, and many children refer to it as a flute, but it is distinctly different. It is blown like a whistle. The bore is cylindrical, and the open end is flared. The recorders used by adults are made of wood, but the type used in elementary schools is usually made of plastic.

The Clarinet

The clarinet is one of the most popular of the reed instruments. It uses a single reed, which is usually made of cane (see fig. 100). Its bore is uniform in size and ends with a flared open end. The holes in a clarinet may be covered by fingers or keys; it has many more than the six holes needed for a diatonic scale; in fact, it usually has 17. Most of the additional holes are needed for sharps and flats.

Clarinets come in several keys, the most common being E-flat and B-flat, both of which are in the soprano range. The range of the E-flat clarinet is from G_3 to $G^{\flat}_6$, and the range of the B-flat clarinet is from D_3 to F_6. There is also a bass clarinet that looks more like a saxophone than a clarinet. The soprano clarinet has a length of 65 cm, and the bass clarinet is 95 cm long.

Clarinets are made from a variety of materials, including wood, plastic, hard rubber, metal, and ivory. African hardwood is particularly popular for professional models. Cheaper models are made of plastic. Hard rubber such as ebonite was used early on, but it is not common today.

The clarinet is used extensively in jazz bands. It was particularly popular throughout the jazz and big band eras and is still quite popular today. It is also an important instrument in the classical orchestra, and many well-known composers wrote for it, including Mozart, Copland, Strauss, and Stravinsky.

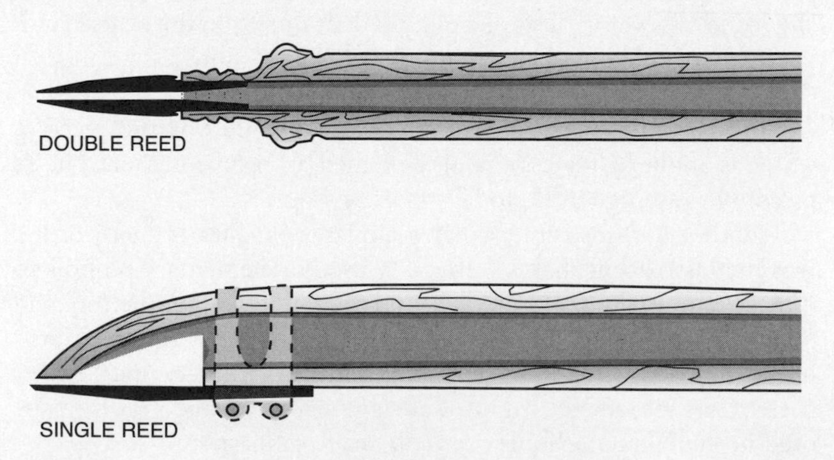

Fig. 100. Single reed and double reed mouthpieces.

The Saxophone

With its long, tapered tube and curled bore, the saxophone is easy to recognize. Patented by Adolph Sax of France in 1846, most models look like a giant curled pipe, although the soprano sax is straight and looks like a clarinet.

The saxophone is usually made of brass and has a single reed mouthpiece. It is usually associated with popular music, particularly big bands and jazz and blues, although early on, it was used extensively in military bands. It has a range of two and a half octaves and has 21 to 23 keys depending on the type. Some of the more common saxophones are the B-flat soprano, E-flat alto, B-flat tenor, and B-flat bass. The most familiar is the tenor saxophone, which is about 88 cm long.

The Other Reed Woodwinds: Oboe and Bassoon

In addition to the clarinet and saxophone there are several other reed instruments. Among them are the oboe and the bassoon. The oboe, a double reed instrument (see fig. 100), is the soprano of the woodwinds; it has a range from $B^{\flat}_3$ to A_6 and is about 60 cm long. It is commonly made of wood with nickel-silver keys.

The bassoon is also a double reed instrument; its range is from $B^{\flat}_1$ to $E^{\flat}_5$. It is generally considered to be the most complicated of the woodwinds. It has relatively complicated fingering, making it a difficult instrument to learn. The mouthpiece is also different—it sticks out the side of the instrument on a narrow tube. The bassoon is usually made of wood, maple in particular, and it has a conical bore.

The Woodwind Virtuosos

We had an introduction to Benny Goodman at the beginning of the chapter. Many still consider him to be one of the greatest clarinetists of all time. Born in 1909 in Chicago, he began taking lessons at the age of 10; with a natural inborn talent, he made rapid progress and was soon playing professionally. He was strongly influenced by New Orleans jazz, and it played an important role in his music throughout his life. At 16 he joined the Ben Pollack Orchestra in Chicago, which at the time was one of the top bands in the United States. He was soon making recordings, and it wasn't long before he formed his own band.

Although Goodman was relatively well known before 1935, it was the change in his style that occurred in the Palomar Ballroom in Los

Angeles that really caused his career to take off. He soon became known as the "King of Swing." And a few years later—in 1938—he was playing in Carnegie Hall in New York City; at the time, this was something new for a jazz orchestra. The concert was a tremendous success; indeed, it was regarded by some as a turning point in jazz. After years of appealing only to specialized audiences, jazz had finally broken through and was being accepted by mainstream audiences.

Another great jazz clarinetist was Artie Shaw, who was born in 1910 in New York City. Shaw began by playing the saxophone, but at 16 he turned to the clarinet and soon left home to tour with a band. For the next 10 years he performed with several different bands; then he formed his own band. Some of the songs he made famous were "Begin the Beguine," "Moonglow," and "Stardust." Shaw signed the African American singer Billie Holliday as a vocalist in 1938, becoming the first bandleader to sign a black female singer full-time. The Artie Shaw Band played to troops in World War II, and like Goodman's band, it also performed at Carnegie Hall.

Shaw was almost as famous for his personal life as he was for his music. He was married eight times; two of his wives were the actresses Lana Turner and Ava Gardner. He was also married to Kathleen Winsor, the daughter of composer Jerome Kern and the author of the best-selling romance novel *Forever Amber* (highly controversial at the time for its frank sex scenes). Quite an assortment, and he said that he never got along with any of them. He also appeared briefly in several movies and in later life wrote several books of fiction.

Two other well-known clarinetists were Pete Fountain and Woody Herman. Born in New Orleans, Fountain played for a while with the Lawrence Welk band but returned to New Orleans, formed his own band, and also bought a jazz club. He recorded numerous records and CDs and is usually associated with Dixieland and jazz. Herman was born in Wisconsin and is closely associated with the blues. One of his most famous pieces is the "Woodchopper's Ball," but he is also well known for "Blues in the Night." Finally I should mention the clarinetist Acker Bilk, who made a big hit of "Stranger on the Shore."

What about famous saxophonists? They aren't as well-known, but there were a few of them. Three of the best known are Charlie "Bird" Parker, John Coltrane, and Stan Getz. Parker was born in Kansas in 1920 and is best known for his contributions to bebop. Much of the time he worked with trumpeter Dizzy Gillespie.

12

The Most Versatile Instrument

The Singing Voice

On September 9, 1956, he stepped onto the stage of the *Ed Sullivan Show* in New York. Within minutes the young girls in the audience were screaming, and they continued screaming until he left. And with the largest TV audience up to that time—estimated to be over 60 million—there no doubt was some screaming in front of TV sets as well. Who was the big attraction? Figure 101 gives it away: Elvis Presley, a singer who had been an unknown only a few months earlier. He was already on his way to stardom, but with that show he was soon a national phenomenon. Girls screamed at all his shows, and at many of the shows there were riots. His accompanist, Scotty Moore, said, "He'd start out with 'You Ain't Nothing But a Hound Dog' and the fans would go to pieces. It happened every time."

At the time Elvis had had few voice lessons—his family was so poor they couldn't afford to pay for any—but he had been singing for a number of years. He had received a guitar for his eleventh birthday and practiced it religiously in his basement laundry room, but the only place he got noticed was in his 1952 high school variety show. He won first place. After graduation he began working as a truck driver.

In 1953, curious to hear what his voice sounded like on a recording, he went to the Sun Records Company to make a recording. He sang "My Happiness" for one side and "That's When Your Heartaches

Fig. 101. Elvis Presley.

Begin" for the flip side. He gave the record to his mother as a birthday present. A few months later, in January 1954, he went back to Sun Records to record two more songs, and these came to the attention of Sam Phillips, owner of the company, who was also a talent scout. A few months later while Phillips was in Nashville, he acquired a demo record of the song "Without Love" and decided to record it at his studio. He couldn't locate the artist who had sung the demo, however, and wasn't sure what to do. His assistant suggested that they use the "truck driver" who had recorded some songs earlier, and he agreed. She phoned Elvis and he came in and recorded the song. While Phillips was not impressed with his rendition of it, he nevertheless asked him if he knew any other songs. Elvis decided to sing a blues song entitled "That's All Right." This time Phillips was impressed; he recorded it and took it to the DJ at the local radio station. It quickly became a local hit. Within a short time Phillips had over 5,000 orders for the record. And as the saying goes, "the rest is history."

I never heard Elvis in person, although he did perform in the city where I was living shortly after he became famous. I delayed and couldn't get tickets, but he was certainly the talk of the town at the time. I have, however, heard many other singers over the years and have enjoyed all of them. Lately, there has been a tremendous amount of interest in the singers on TV's *American Idol.* I'll have to admit I'm a fan of the show. It's interesting to listen to the singers and compare all the different voices, and of course, make a guess at who is likely to win.

What is it that makes one voice stand out above others? What makes a voice pleasing to listen to? What was it about Elvis's voice that made him such a success? Why were so many people attracted to Frank Sinatra's voice? And why have opera fans been so enchanted with Luciano Pavarotti's and Plácido Domingo's voices? In this chapter we'll look at how the singing voice is created and why some are more pleasant to listen to than others. It's still a mystery why some singers with rather average voices become instant celebrities, while others with good voices get nowhere. No doubt many things are involved in musical success, but a relatively good singing voice is still essential. No one will argue about that.

Singing through History

Singing is the oldest form of music. Even though simple instruments can be traced back thousands of years, there's little doubt that people were singing well before the first of these instruments appeared. So it's safe to say that the voice is the oldest of the musical instruments (assuming, of course, that we can call it an instrument). We know that singing played an important role in the ancient world; the Jewish psalms were songs, and Greek drama used singers. Early in the current era most public singing was performed in Christian churches, and almost all of it was done by males. Strangely, the female voice was not "discovered" for many years. Much of the early church singing was in the form of chants; in fact, as early as 600 Pope Gregory set up schools to teach chant, and the technique eventually became known as Gregorian chant.

The church continued as the center for singing for many years. At first most singers were tenors, but gradually alto and bass voices were added. In the high range young boys were used, but what are called "falsetto" male voices (we will discuss them later) were also used. By the fifteenth and early sixteenth centuries, however, women finally began to take part, and soon sopranos began to play a large role. By the eighteenth century singing really began to come into its own; it spread beyond the church pews and became increasingly popular in public gatherings and in public buildings. Singing schools were established, and many books on singing were published.

This was also the era when instrument development was at its peak. Musicians were continually striving for a clear, bright, beautiful tone, and the same challenge also began to apply to the voice. The audience for music was now increasing rapidly, and larger and larger music halls were being built. Opera was becoming popular, and many of the well-known composers were writing operas. Some of the earliest operas were written by George Handel in England, but it wasn't long before the scene switched to Germany and Italy. Mozart wrote several operas, including *The Magic Flute* and *The Marriage of Figaro;* and in Italy, Giuseppe Verdi wrote *Rigoletto* and *La Traviata.* The peak, however, came with Richard Wagner in Germany; three of his most famous operas were *Tristan and Isolde*, *Tannhäuser*, and *Lohengrin.* Richard Strauss also wrote a number of important operas.

One of the problems associated with opera is that the singer's voice has to stand out above the orchestra. For untrained singers this is difficult, if not impossible. Highly trained singers, however, are taught (as are stage actors and sometimes speakers) to "project" their voices to the audience with greater volume. With the twentieth century came popular music, and the problem of projection was overcome with the invention of the microphone. Many different types of singers soon emerged, including jazz singers, folk singers, and "crooners" such as Bing Crosby. Modern singers now use many electronic devices to amplify sound.

Let's turn now to what makes the singing voice of some people so enjoyable.

The Voice Organ

Several parts of the human anatomy are involved in singing, so let's begin with them. Most of them are illustrated in figure 102. First of all we have the lungs; they are the powerhouse behind the voice. Next we have the larynx, or voice box; it leads to the pharynx, or throat tract. The mouth, nose cavity, and lips also play an important role in singing, as do the vocal cords and the opening through them, called the glottis. Within the mouth are the hard and soft palates, and finally there is the epiglottis, a flaplike valve on the top of the lar-

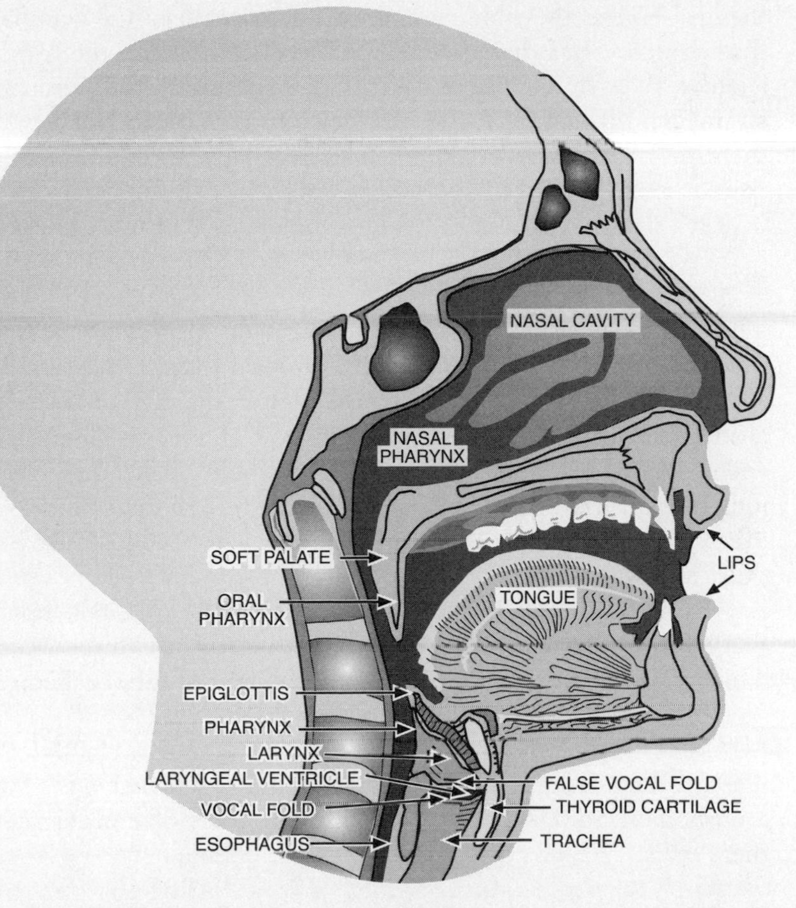

Fig. 102. Cross section of the mouth and throat showing the parts of our anatomy important for singing.

ynx that drops down when we swallow to prevent food from entering the trachea.

We can lump these anatomical parts into three major units that are particularly important for a singer. They are

1. a power supply (the lungs),
2. an oscillator or vibrator (the vocal cords), and
3. a resonator (the vocal tract).

These three units are also common to most musical instruments. For example, in the case of the trumpet the lungs are the power supply, the lips serve as the vibrator, and the air chamber of the trumpet is the resonator. But there is a basic difference between the trumpet and the voice. In the case of the trumpet, and most other similar instruments, we have *feedback* which helps form the standing wave. In the case of the voice we do not have feedback. In this respect the voice is more like a reed instrument, such as a harmonica.

Let's turn now to a more detailed look at each of these units.

The Lungs

The major function of the lungs is to provide air pressure; it is this pressure that causes the vocal cords, or vocal folds, to vibrate. The air from the lungs passes through the glottis (fig. 103), and it is the glottis that cuts it off. The lungs usually hold about 3 to 4 liters of air, and when we breathe we inhale and exhale about a half a liter. Adults' lungs can, of course, hold more than this in certain cases; if you take a deep breath and hold it, for example, there will be about 5 to 6 liters of air in your lungs.

The air in your lungs is expelled through a tube called the trachea that leads to the vocal tract, or more particularly, to the vocal cords. This air passes through the glottis in the center of the vocal cords.

The Oscillator

The vocal cords act as an oscillator or vibrator. The name is a bit of a misnomer, however, as they are actually folds of mucous tissue; there are, in fact, no "cords" associated with them, but there is carti-

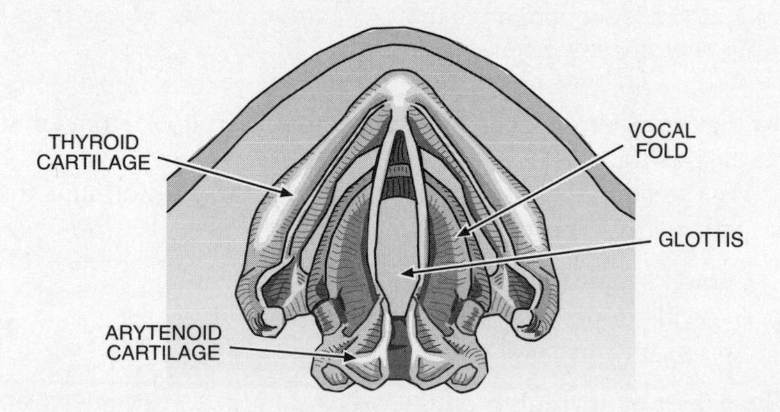

Fig. 103. The vocal cords (folds).

lage within them. Because of this, some people like to refer to them as vocal folds; most of the time, I'll use the more common term "vocal cords," but I will occasionally refer to the folds.

The vocal cords, shown in figure 103, are attached in the front and move apart at the back. They are closed for swallowing. The folds are controlled by many muscles, and these muscles change the shape of the opening. The two front folds meet at the Adam's apple and are attached to it at the front of the throat. The back ends, which are quite mobile, are attached to cartilage called the arytenoid cartilages. For low notes the opening of the cords is fairly wide, but as the sides are stretched and come closer together, the pitch increases. For the very highest notes the opening is long and narrow, and the stretching is at its greatest.

So how do the vocal cords vibrate? When you speak, sing, or make a noise, the air from the lungs hits the cords and applies a pressure. This pressure causes the vocal cords to open, and a small puff of air passes through. The muscles controlling the opening, however, quickly close, and the air is cut off. But because you are still making the sound, the pressure from the lungs is still there, and as a result the cords open again. This opening and closing is very much like the opening and closing of the lips of a trumpet player. The vocal cords are, in fact, vibrating, and as we will see, many different frequencies are associated with this vibration.

The frequency of the vibration is determined by the air pressure of the lungs and by the mechanical properties of the vocal cords, which in turn are regulated by the surrounding muscles. If you could listen to the noise they make, it would sound like a buzz, similar to the buzz you get when you blow on a trumpet mouthpiece.

For normal conversation the vibration range for males and females is as follows:

Males: 70–200 Hz
Females: 140–400 Hz

These ranges of vibrations change significantly for the singing voice and are generally much higher. The frequencies that are generated have a characteristic spectrum consisting of the fundamental and a large number of overtones. The amplitude, or loudness, of the overtones decreases uniformly with frequency at the rate of about 12 dB per octave. (You can see this in figure 104.) The vocal cords are the

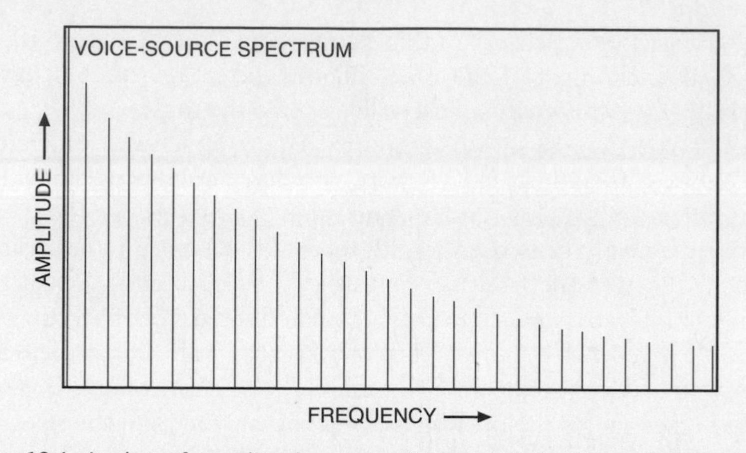

Fig. 104. A plot of amplitude versus frequency for the voice.

"voice source" for both ordinary speech and singing, and surprisingly it is quite similar for both singers and nonsingers.

Earlier, I mentioned the "falsetto" voice that male singers can produce; these vibrations are much higher in frequency than normal tones. These tones are produced in the larynx by a combination of two effects. The first is related to the shape of the glottis as air passes through it; the opening is elliptical in this case, and as a result the folds do not fully close. In addition, the muscles controlling the folds allow the edges to become thinner and vibrate more easily.

So, now that we know how the vibrations are created, let's consider what happens to them as they enter the resonator.

The Resonator

For the human voice, the resonator is a chamber that we refer to as the vocal tract. It plays the same role as the air chamber in the trumpet, and consists of the larynx, the pharynx, the mouth, and sometimes the nasal chamber. Like any chamber it has certain resonances, or frequencies, that allow sounds to pass through with much greater amplitude. Sounds in this frequency region are therefore much louder than those in other regions. This difference is quite significant in the case of a solid cylinder, but the vocal tract does not have distinct resonances, mainly because the walls are soft and absorb sound waves much more than a solid wall does. Because of this, the resonant frequencies are not sharp and distinct. There is a peak, but the frequencies close to the peak also have considerable amplitude, so what we have is actually a group of closely associated resonances. We call them *formants.*

The formants depend on the shape of the vocal tract, and their position can be changed by changing the form or shape of the vocal tract. We usually refer to the fundamental and first few overtones as the first, second, third, and so on, formants. Normally there are about

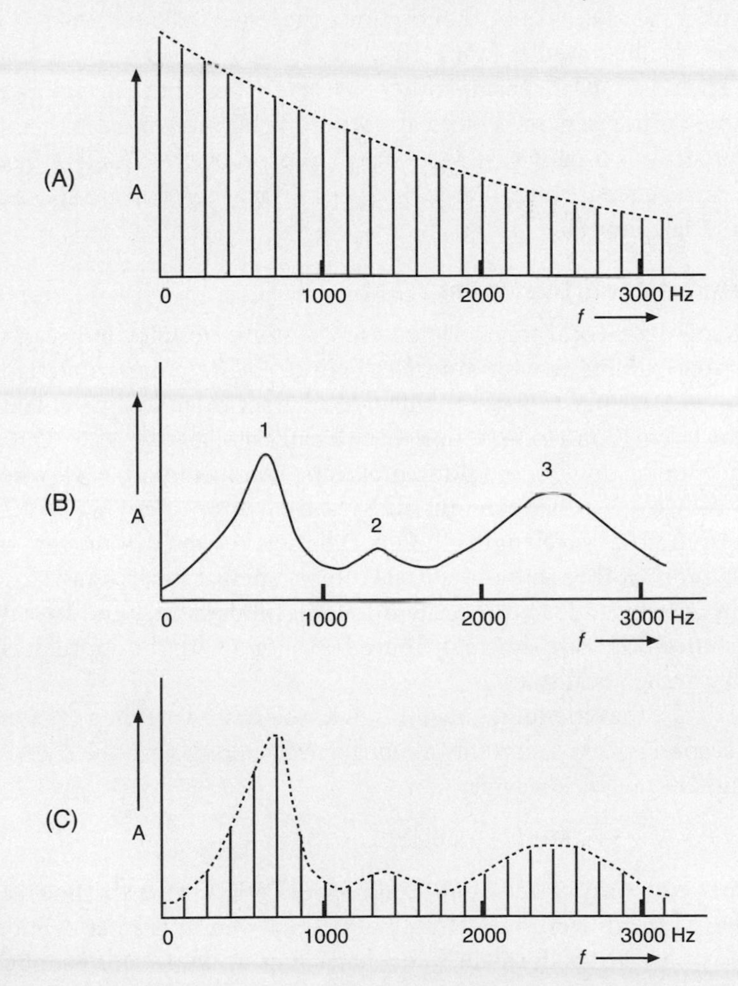

Fig. 105. Formants for the human voice: (A) the spectrum of sound frequencies as it leaves the vocal folds; (B) the effect of filtering due to the frequency response of the vocal tract; (C) a superposition of the frequencies.

five. Studies have shown that the first formant is sensitive to the jaw opening; in particular, it contracts the vocal tract near the glottis and expands it near the lips. The second formant is sensitive to the body of the tongue and the third to the position of the tip of the tongue. So we obviously have some control over them.

As the voice source passes through the resonator, its frequency spectrum will be affected by the formants. There will be a dropoff in intensity, but because of the formants there will also be peaks. The process is illustrated in figure 105.

The position, shape, and other characteristics of the formants are unique to the person. Voices are different because vocal tracts are different. It is usually easy to identify a person by the sound of his or her voice, even if you can't see that person (as when you are speaking on the telephone).

Formants and the Vocal Tract

Although the vocal tract changes and is not exactly cylindrical, we can approximate it with a perfect cylinder. The comparison is far from perfect, but it is useful. The vocal tract in an average male is about 17 cm (7 in.) long. Using this we can calculate the approximate frequency of the first few formants using the formula $v = \lambda f$, where v is velocity, λ is wavelength, and f is frequency. First, we have to determine the wavelength, λ. Our cylinder is open at one end and closed at the other, just as the vocal tract is open at the lips and closed at the glottis, and we can easily draw the fundamental and first few overtones. They are shown in figure 106, along with the approximate nodes in the vocal tract.

We see that the fundamental is $\frac{1}{4}\lambda$, the first overtone is $\frac{3}{4}\lambda$ and the second is $\frac{5}{4}\lambda$. Using this in our formula, along with the speed of sound (340 m/sec), we get

$$f = v/\lambda = 340/4(0.17) = 500 \text{ Hz}.$$

In the same way we get 1,500, 2,500, and 3,500 Hz for the first, second, and third overtones. If the vocal tract were a perfect cylinder, these would be the formant frequencies, but we know this is not the case; nevertheless, these values are relatively close. Furthermore, as we saw, the formant frequencies can be changed considerably. In a male, for example, the first formant can be changed from 250 to about

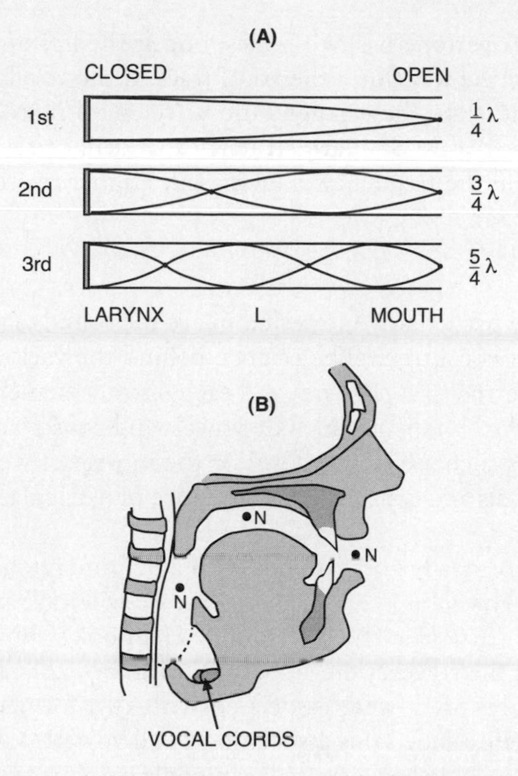

Fig. 106. Waves in the vocal tract: (A) the fundamental and first two overtones; (B) the position of the nodes (N) in the mouth.

700 Hz. In the same way, the second formant can be changed from about 700 to 2,500 Hz. And higher formants can also be changed.

Let's turn now to how the vocal tract produces these different resonant frequencies. Each, of course, corresponds to a standing wave, and it is well known that certain points along a standing wave, in particular, the nodes, are sensitive to changes in the tube width. If the throat is contracted or expanded near the nodes, that change in width changes the entire pattern. A contraction, for example, stretches the pattern out and increases the wavelength, while an expansion shortens it and decreases the wavelength. Looking at the waves more closely, we know that the node of the fundamental is at the lips, so if we widen the lips (open the mouth wider), the wavelength will be decreased and the frequency increased.

The first overtone has two nodes: one at the lips and one about two-thirds of the way down the vocal tract. If the vocal tract is constricted at either of these points, the wavelength increases and the frequency decreases. Similarly, if it is expanded, the wavelength is shortened and the frequency is increased. Higher overtones are affected in the same way.

Phonemes

What we are particularly interested in are the sounds that we finally get from the vocal tract. The science behind the various sounds of speech is referred to as *phonetics*, and each distinct element is referred to as a *phoneme*. Phonemes include vowels and consonants as well as other sounds such as semivowels. For the most part we will discuss only the vowels and consonants, as they are of particular importance in singing.

Vowels are steady vocal sounds with a definite pitch. We usually think of the vowels as *a*, *e*, *i*, *o*, and *u*, but we will look at them more generally as particular types of sound. Helmholtz showed in 1860 that some of the vowels were associated with formants; in particular, he showed that they were associated with two formants. In 1924 Robert Paget extended this association to all vowels, and this meant that all vowels could be generated using only two formants. We know now that other formants are also associated with them to a lesser degree, but we'll ignore them for now.

Figure 107 shows the formant frequencies associated with various vowels, or vowel sounds. The graph gives us the formant frequencies for the vowel sounds *ee*, *aa*, and *oo*. Looking at *aa* (in *hard*) we see that the first formant is about 570 Hz, and the second is about 1,100 Hz. The first formant is of high frequency; for this we need a wider mouth opening, according to the last section. For *ee* (*heed*) the first formant frequency is low, so we need a more closed mouth opening. And so on for other vowels; we will talk more about them in the next section.

Consonants are quite different from vowels and are sounded in a number of different ways. They are divided into several types, according to how they are formed. *P* and *t* are called plosives, *f* and *v* are fricatives, *m* and *n* are nasals, and *w*, *r*, and *y* are semivowels. The first of the plosives, *p*, is made by closing your lips and suddenly forcing them apart. The second, *t*, is made in the same way, but with the

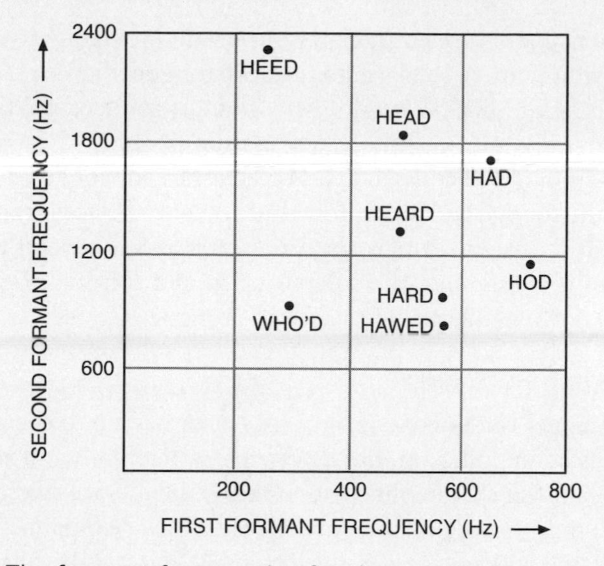

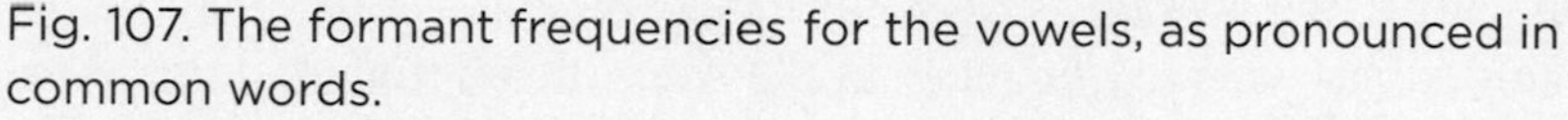

Fig. 107. The formant frequencies for the vowels, as pronounced in common words.

tip of your tongue against the roof of your mouth. The nasal consonants *m* and *n* are sounded in the nasal cavity, with most of the sound coming out of the nose. Others are formed similarly.

The Articulators

As the above discussion indicates, several parts of the facial and throat anatomy are critical in forming the various sounds, including vowels and consonants. We refer to these anatomical parts as articulators, and we have, of course, discussed several of them already, but let's look at them in a little more detail. The major articulators are the jaw, the tongue, the lips, and the larynx. The movement of the articulators changes the structure of the vocal tract in certain ways. One way is to constrict or dilate it, and this, in turn, as we saw, changes the position of the formants. And the changing position of the formants has an effect on the sound that is emitted, and in particular, on the vowels. The most sensitive regions, as we saw, are the nodes.

The first formant is especially affected by the jaw opening, the second by the body of the tongue, and the third by the placement of the tongue. Remember also that men and women have different sized vocal tracts and have different formant frequencies.

The articulators are particularly important in relation to the vowels and consonants. The placement of the tongue, the soft palate and the shape of the lips all modify the sound produced by the vocal cords. Singing requires perfect articulation of the vowels and good pronunciation of the consonants. We have already looked at the position of the tongue and lips in forming some of the consonants. The position of the articulators is also important in the case of the vowels. For *ee*, for example, the tongue is up and forward. For *aa* the tongue is down and back, and for *oo* the tongue is lower.

The Singing Formant

An opera singer is trained to project his or her voice to be heard above the sounds of an orchestra, but the art of projection is a problem for most other singers (overcome, of course, by the use of a microphone). The reason can be seen if you analyze the voice frequencies of most singers. Their average is typically about 450 Hz, and this is about the same as the average frequency of the sound from the orchestra. Furthermore, in both cases there is a sharp cutoff above about 2,000 Hz. Thus, it would appear that the sounds of an orchestra would drown out a singer's voice. But in the case of a trained opera singer they don't. In fact, she doesn't even need (or use) a microphone. Why? The reason is that a trained opera singer has a distinct formant peak at much higher frequencies than the usual peaks—usually around 3,500 Hz—and it can easily be heard above the orchestra. It is referred to as the *singing formant*. It is, in fact, at an optimal frequency: high enough so it is well above the orchestra's frequency range, but not too high for the singer to lose control of it. Furthermore, singing at this frequency range requires no extra vocal effort on the part of the singer.

This "extra" formant is accomplished by lowering the larynx. When it is lowered sufficiently, it has its own standing wave and therefore a resonance of its own, one that is much higher in frequency than the other resonances of the vocal tract.

Characteristics of a Good Singing Voice

It doesn't take an expert to spot a good singing voice; almost anybody can do it. And many of us have fantasized at one time or another about becoming a famous singer (usually when we're singing in the shower). But what are the main characteristics of a good singing voice? Let's begin by looking at the difference between the singing and speaking voice. The singing voice is usually described as being "darker" than the speaking voice. Something like this darker voice is produced when a person yawns and talks at the same time. When someone sings, the larynx is lowered and the lowest part of the pharynx is expanded.

If we look more closely at this difference, we see that the speaking and the singing voice differ considerably in their wave structure at different frequencies. A large number of frequencies are present in speech; when they are filtered by the resonances of the vocal tract, the resulting signal still looks like the frequency response curve of the vocal cords because the frequencies generally "fill out" the formants. When a note is sung, however, there may be only a few frequencies, and they may not fill out the formants nearly as much. This is particularly true at higher frequencies, where the formants may be widely separated. Singers can, however, compensate for this to some degree.

Turning now to the features of a good singing voice, we find that control of the position or height of the larynx is critical. Most singers may not be able to control the larynx the way opera singers can, but if they are good singers, they should have a fair amount of control over it. When we talk, we are not aware of raising and lowering the larynx but we do it continually. Good singers have to have a much greater degree of control over the position of the larynx.

In addition, a good singer has to have a relatively large vocal range and—of prime importance—a voice with a pleasing tone. Range and tone depend to a large degree on the shape, length, and other characteristics of the vocal tract and on the vocal cords. A person's range of pitch is also set by the size, shape, and symmetry of the vocal cords, and the strength and control of the chest and vocal muscles. In addition, the number and structure of the overtones that a singer produces is of critical importance.

Singers are usually divided into the following groups according to their range:

Soprano	C_4 to C_6	(264–1,047 Hz)
Alto	G_3 to F_5	(196–698 Hz)
Tenor	D_3 to C_5	(147–523 Hz)
Baritone	A_2 to G_4	(110–392 Hz)
Bass	E_2 to D_4	(82–294 Hz)

Most pop singers are tenors; indeed, most singers in other genres besides opera are also tenors. Generally, the other voice parts are of importance only in opera singing and in choruses or choirs.

Two other features of a good singing voice are vibrato and tremolo. Vibrato is the periodic change in the frequency of a note, usually by about 5 to 10 Hz. The period is usually very short, and the loudness remains constant during the vibrato. Tremolo, on the other hand, is a periodic change in loudness without changing the frequency. It is produced by pulsations in the muscles of the larynx. In many cases the two effects come together.

Famous Singers: A Few Who Made It

We had a brief introduction to the career of Elvis Presley at the beginning of the chapter. Let's turn now to a number of other singers. One of the most famous twentieth-century singers was Frank Sinatra. Born in Hoboken, New Jersey, into a middle-class family, he began his career during the swing era. His first job was with Harry James's band, where he recorded several songs and gained considerable experience. Within a year he left to join the larger Tommy Dorsey Orchestra and soon had a hit with "I'll Never Smile Again." Over the next few years he became extremely popular with "bobby soxers"—young girls who, in the fashion of the 1940s, wore skirts and white socks, and who swooned and screamed when he sang. By 1943 Sinatra had signed with Columbia Records, and soon he was the second most popular singer in the country, after Bing Crosby. He was so popular that between 1940 and 1943 he had 23 top-ten singles.

In 1945 he began combining singing with acting, when he starred with Gene Kelly in *Anchors Aweigh* and *Take Me Out to the Ball Game*. By 1948, however, his career was starting to stall, and sales of his records were declining. They continued to decline during the early 1950s, and in 1952 Columbia Records dropped him. It was about this time that he married the actress Ava Gardner. While his career continued to decline, hers was rising; this situation had a serious effect on the marriage—it ended a couple of years later.

In 1953, things finally began to turn around for Sinatra, and he was soon on his way to a comeback. It began when he won an Academy Award for best supporting actor for his part in *From Here to Eternity*, and continued with several hits over the next few years and with starring roles in *Not as a Stranger* with Robert Mitchum, and *The*

Tender Trap with Debbie Reynolds. Then in 1955 he hit it big with the movie *The Man with the Golden Arm.* Over the next few years he had hits with songs such as "Come Fly with Me," "I Could Have Danced All Night," and "Dancing in the Dark." During these years he spent much of his time in Las Vegas, where he became associated with the "Rat Pack," a group consisting of Sinatra, Dean Martin, Sammy Davis Jr., Peter Lawford, and Joey Bishop. They were a highly popular group.

In later years Sinatra had hits such as "Strangers in the Night," "That's Life," and "My Way," a French song translated by Paul Anka that became identified with Sinatra.

Sinatra had a free-flowing vocal style along with a legato (smooth) style of singing and phrasing. He also had a relatively large range, extending from high F down to low E. He had excellent phrasing and an excellent ability to hold long notes.

Another singer with a long and successful career is Tony Bennett. Born in Queens in New York City, he grew up listening to Al Jolson, Bing Crosby, and jazz singers such as Louis Armstrong. He particularly enjoyed jazz, and many of his songs have been "jazzy," but he is usually thought of as a pop singer. Bennett was drafted into the army in 1944 and saw considerable action in the last months of World War II (in fact, he was almost killed). After the war he decided to continue with his singing career. Even though he had been singing for years by then, he was still an unknown. He was finally noticed by Pearl Bailey in 1949 and was soon her opening act. In 1950 he signed with Columbia Records.

His first big hit was "Because of You," which reached number one on the hit parade in 1951. It was followed by a new rendition of Hank Williams's "Cold, Cold Heart," and in 1953 he had a smash hit with "Rags to Riches," which topped the charts for eight weeks. But in 1955 the rock and roll era began, and Bennett's career, along with those of the other "crooners," began to decline. One of the few hits he had in the late fifties was "In the Middle of an Island," which reached number nine on the charts.

The song that is most identified with him today, "I Left My Heart in San Francisco," came out in 1962. It reached only number 19 on the pop charts, but his album of the same name was in the top five, and the song won a Grammy award for him. This was his last success for many years. In the mid-1980s, however, he made a comeback, not with new music, but mainly with his old songs. In a sense he "discovered" a new audience for his music, and to his delight, many in that audience were young. He is still touring and singing in his eighties.

Bennett has a pleasing tenor voice with an unusual way of phrasing. Early on, he learned to imitate various instruments in the band, and later used it occasionally, much to the delight of the audience.

There are, of course, many other important popular singers worthy of mention, but all I can do is name a few. One of the early greats of jazz was Billie Holiday, and of course Louis Armstrong was as famous for his raspy voice as he was for his trumpet. Other popular singers have included Bing Crosby, Dean Martin, and Ray Charles, and later, Whitney Houston, Barbra Streisand, Madonna, Celine Dion, Billy Joel, and Elton John. In country music, important early artists were Jim Reeves, Patsy Cline, and Hank Williams; later singers have included Johnny Cash, Garth Brooks, Reba McEntire, Alan Jackson, and Vince Gill.

On the classical side there have been a large number of important singers, and again all I can do is mention a few. One of the greatest early opera singers was the Italian Enrico Caruso. He was, in fact, the most popular singer of any genre in the early 1900s. Born in Naples, Italy, he was one of seven children in a poor family. At 18, by singing at a local resort, he was able to buy his first pair of shoes. He first came to the United States in 1903, when he made his debut at the Metropolitan Opera in New York City. Three years later he was performing *Carmen* in San Francisco; in the early morning hours after his performance, the Great Quake of 1906 hit, and he barely escaped. He vowed never to perform in San Francisco again, and he never did.

Caruso died at the relatively young age of 48. He had a powerful voice that was known for its range and beauty. Early on he was a baritone but later became a tenor. He was one of the first artists to record extensively. His life was portrayed by Mario Lanza in the movie *The Great Caruso* in 1951.

One of the best-known modern opera singers was the tenor Luciano Pavarotti, who died in 2007. Born in Milan, Italy, Pavarotti made his American debut in 1965 at the Miami Opera. This performance was only a moderate success, but in 1972 his appearance at the Metropolitan Opera in New York easily made up for it. He received 17 curtain calls at the end of the program—an all-time record number. Over the next few years he appeared at opera houses around the world and made numerous television appearances. He also won several Grammies and received gold records.

Pavarotti was one of the highly successful "Three Tenors," which consisted of Pavarotti, Plácido Domingo, and José Carreras. They first sang together at the World Cup finals in Rome in 1990, and went on to numerous other appearances, both on television and live.

In the 1990s Pavarotti set an attendance record at Hyde Park in London with an outside concert attended by 150,000 people. In 1993 he topped this with a concert in New York's Central Park that was attended by 500,000 people.

Also world famous is Plácido Domingo, one of the first well-known opera singers not from Italy. He was born in Madrid, Spain, but moved to Mexico at the age of eight. He studied music at the National Conservatory of Mexico. In 1966 he made his debut in the United States at the New York City Opera, and in 1968 he performed at the Metropolitan Opera. Like Pavarotti, he was one of the Three Tenors, and he also had an extensive career, singing 92 different roles on stage, mostly at major opera houses. He also occasionally diverged into popular music, recording "Perhaps Love" with pop singer John Denver, and he appeared on TV with Julie Andrews. He is known for his versatile, strong tenor voice.

Among female opera stars we have Maria Callas, Beverly Sills, and Roberta Peters. Callas was born in New York but received her music training in Athens, Greece. She appeared first in minor roles in the Greek National Opera, and she made her professional debut in 1942 in Greece. In 1945 she left Greece to come to the United States, but soon left and went to Italy, where she established her career. She debuted in London in 1952, and in the United States in 1954. She was a mezzo-soprano with a large range of three octaves who could sing long, difficult pieces with ease.

The soprano Beverly Sills was born in Brooklyn in 1929 and died in 2007. When she was 10 years old, she told her father she wanted to be an opera star. His reaction was surprise and dismay, and he no doubt quickly dismissed the idea. But indeed it came to pass. In later life Sills once said, "You may be disappointed if you fail, but you are doomed if you don't try."

Sills' early career was entirely in the United States. Not until later did she sing in Europe and other places around the world. She made her debut in 1945 with a Gilbert and Sullivan touring company. Her first stage appearance was in 1947 at the Philadelphia Civic Opera, where she played in *Carmen.* During the 1960s and '70s she was one of the best-known opera singers in the world and appeared on the cover of *Time* magazine. After retiring in 1980 she became general manager of the New York City Opera.

Another American opera singer of note is Roberta Peters. Born in the Bronx in New York in 1930, she dreamed of becoming an opera star from an early age. At 13 she sang for opera star Jan Peerce, who was impressed with her voice. At 20 she made her debut at the Metropolitan Opera and was an immediate success. Over the years she has sung numerous operas and appeared in two movies. She was on the *Ed Sullivan Show* a record 65 times.

NEW TECHNOLOGIES & ACOUSTICS

IV

13

Electronic Music

Imagine the following: Four guys decide to form a band. Every evening they practice in the garage of one of the members. Finally they are good enough to start playing at some of the local clubs. Then one day one of them says, "Let's make a recording," and they all agree that it is a good idea. They buy what they need, make a CD, and soon have a hit. This is imagined, of course, but it has no doubt happened hundreds of times across America. And in many ways it has become easier in recent years, particularly with the advent of the Internet as well as the new technology that is now accessible to all musicians.

In this chapter we look at much of this new technology and how it has changed music. Two innovations that have had a tremendous effect are the *synthesizer* and the *sequencer*. As its name suggests, a synthesizer is a device that synthesizes music; in other words, it can create (from scratch) the sound of any instrument from a concert piano to a trumpet or a drum. It might seem that this device would make the musician obsolete, but of course it hasn't. (A few might argue with me here.) The second of the two, the sequencer, is a device that allows musicians to record their music, then edit and make changes to it, and finally, play it back.

Synthesizers and Sequencers

The synthesizer is a device capable of generating electronic waveforms that are used in the creation of musical sounds. There are two types of synthesizers: analog and digital. Analog synthesizers use "subtractive" synthesis to generate their signals; digital synthesizers use "additive" synthesis. Electronic organs are good examples of instruments that use additive synthesis. The electronic systems within them generate sine waves, similar to the ones we discussed earlier, that are mixed or added together to form the complex waveforms that we eventually hear as music. Analog synthesizers use a technique similar to this, but they use subtractive synthesis rather than additive. The process begins with sine waves and other basic waves such as sawtooth and square waves, generated using electronic oscillators. The resulting waveforms are then passed through filters that subtract out various frequencies until the final required waveform is produced.

Electronic oscillators are devices that use electrical circuits to produce various types of waveforms. These waveforms are the "raw" material the synthesizer begins with. The filters through which the waveforms are passed are of several types. The most common are low-pass filters, which allow only low frequencies to pass through, and high-pass filters, which allow only high frequencies to get through.

All the early synthesizers were analog devices. In this type you are dealing directly with the audio signal; in other words, the signal (waveform) itself is manipulated and stored. With the advent of digital techniques, however, digital synthesizers were soon built, and they were quite different from analog synthesizers. In a digital synthesizer the waveform is represented by digits (as in a digital computer). These digits can be retrieved later and transformed back to a waveform, which can be sent to a speaker that produces music. And, as we will see, digits are much easier and convenient to deal with than complex waveforms.

The first synthesizers appeared in labs and sound studios in the mid-1950s. These analog synthesizers were large and cumbersome to operate, consisting of many electronic components—oscillators, amplifiers, and various types of filters. They also had numerous knobs and buttons, and the performer was also continually manipulating and regulating them. Indeed, the operator had to be much more than just a musician; he had to control the sound using the knobs and buttons, and he had to know what he was doing.

One of the first synthesizers was made by RCA; called the Mark II, it was housed at the Columbia-Princeton Electronic Music Center in New York and was a large electronic system that employed vacuum tubes. A number of other large synthesizers were built in the late 1950s and early 1960s, and they were also housed in studios. By this time, although synthesizers were still analog, vacuum tubes were starting to be replaced with transistors, and electronic devices were becoming smaller.

The first synthesizer that was small enough to be used by musicians was built by Robert Moog. The Moog synthesizer, introduced in 1964, was at first little more than a curiosity, but its importance was soon realized, particularly after some hit records were produced using it. The first of these was the million-seller *Switched-On Bach*, which appeared in 1968; a year or so later the Monkees used a Moog synthesizer to produce one of their albums, and it hit the top of the charts.

The Moog synthesizer was small compared to its predecessors, but it still consisted of several components connected together by cables and was not easy to operate. Improvements and streamlining were needed for it to become popular with musicians, and they came in 1970. The improved version, called the Minimoog, was a single unit with built-in keyboard that was portable and relatively easy to use, and over the next few years it became one of the most popular synthesizers ever built. It was even used to generate the music in the James Bond film *On Her Majesty's Secret Service.*

All early synthesizers were monophonic, which means that they could produce only one tone at a time. In the early 1970s the first synthesizers that could sound two notes at the same time were built. Then in 1976 the first polyphonic synthesizer came on the scene. It was capable of producing several tones at the same time. One of the first of the polyphonic synthesizers was the Prophet-5, which could produce five tones simultaneously.

About this time IBM and Apple began producing personal computers, and the technology associated with them increased rapidly. With transistors (and later, integrated circuits) now having replaced tubes, computers were becoming much more compact, and the advances in the computer industry had a direct effect on the electronic music industry. Synthesizers could now be made much smaller and were soon mass-produced for the public. Two different kinds of synthesizers soon came on the market. The first were canonical synthesizers, which generated the timbre and other characteristics of the note by adding together various overtones of the note. The second group were not actually synthesizers but were closely related in that they also produced musical tones; they were called *samplers.* A sample is a brief recording of the sound of a tone from a particular acoustical instrument. The sampler stores all the acoustic information about the sounds (harmonic content and so on) of a range of tones as they would be played by the instrument. Samplers have been extremely important in the music industry, and I will talk about them in considerably more detail later.

By the late 1970s digital techniques were being used extensively, and most synthesizers soon became digital. In fact, within a few years there were few analog synthesizers on the market. The digital synthesizers used pulses of electricity, and all had a digital control interface (a place for input and output leads) for connecting with other units. But there was a difficulty: each manufacturer developed its own design, and the components from one manufacturer wouldn't fit those of another; as you might expect, this caused problems. Finally, with a push from many musicians and a number of manufacturers, a universal system was presented in 1983. It was called MIDI, which stood for Musical Instrument Digital Interface.

All equipment since then has had MIDI connectors, namely MIDI IN and MIDI OUT. Through these connections, any machine, regardless of its manufacturer, can be linked with any other machine. MIDI is not a device: it is an interface and language for passing messages between different electronic musical instruments so that they can communicate with one another.

When you play an instrument equipped with MIDI, a sequence of MIDI commands is generated that describes everything about the note: its frequency, duration, velocity, and so on. The sound itself is not recorded, only a set of MIDI commands. To play back the sound, a MIDI OUT signal is sent back to the keyboard, or a sound module, where it triggers a sound.

With the development of MIDI came many MIDI devices that

soon played a central role in the music industry. One of them was the MIDI *sequencer*, which of course is the sequencer we have been talking about, but in MIDI form. It is used to record the MIDI data generated by a keyboard, which can then be edited and stored. A sequencer allows the musician to record his performance in the same way as with older cassette tape recorders, but no tape is used; only MIDI messages are recorded. For example, when you play the note G on a keyboard, the sequencer stores digital information corresponding to G. Later you can tell the sequencer to play the recorded note, and it will play G.

About the same time the first programmable rhythm, or drum, machines came on the market. The first appeared in 1978 and were soon followed by several others. The earliest ones were analog, but the first digital machines appeared soon after in 1980. Closely associated with drum machines were samplers. As we saw above, the sampler is an electronic musical instrument similar to the synthesizer. The main difference is that instead of generating musical sounds from scratch the way a synthesizer does, the sampler starts with recordings, or "samples," of various types of musical sounds. Samplers can play back virtually any type of recorded sound, and they also usually contain editing capabilities for altering or modifying the sound. As a result, samples and sampling have become one of the most exciting developments in music in recent years. Most of the early samplers were separate hardware units, but sampling is now frequently done using computers and software. Various other types of external modules are also now used. We will discuss them in more detail later.

Analog versus Digital

As we saw, all early electronic music devices were analog; they treated the sound wave as an electronic signal. But with the rapid development of digital techniques in computers, digital synthesizers soon came on the market (about the mid-1980s), and the sounds that came from them were exact and clinically perfect. Musicians, however, had grown used to the slight imperfections of analog sounds, which sounded more like natural sound to them. Digital sounds seemed to lack the "warmth" of analog sounds. So while analog machines quickly fell out of fashion when digital ones were introduced, many musicians soon began to long for the "old" sounds. As a result, technology from analog machines has now been incorporated in many digital machines creating a kind of hybrid. Indeed, some of the older analog models have actually been resurrected and brought back on the market, though most now use some digital technology.

Setting Up a Recording System

Now that we have some idea of what modern recording systems are like, let's look briefly at the various alternatives for setting up such a system. Since MIDI is now at the center of all systems we usually refer to such systems as MIDI systems. There are, in fact, three different approaches to setting up such a system, but in recent years one of them—software sequencing—has started to overshadow the others. As we will see, it makes extensive use of the computer. Since it is now so important, most of our discussion (particularly in the next chapter) will be directed toward it.

The first of the three approaches for setting up a system is referred to as the *stand-alone* system. This is the type of system that musicians had several years ago (if they had a system). It consists of several separate units: a synthesizer, a recorder (which may be contained within the synthesizer), a mixer (for mixing sounds), monitors, usually a microphone or two for external audio, and several other devices. This is the type of system you see in a large professional studio, and it is particularly good for recording live performances. For most amateurs, however, it is an expensive route to take. Stand-alone sequencers, recorders, mixers, and so on are expensive, and when you add everything together it can add up to a small fortune. Using this kind of recording system also requires a fair amount of technical knowledge and ability. Furthermore, everything has to be connected in just the right way, as all connections are external.

The second of the three recording systems is usually referred to as the *studio-in-a-box* (SIAB) system. In this case, recorders, mixers, effect processors, and so on are all contained in one unit, so that you don't have to worry about external connections. External instruments and microphones can be used with it, but everything you need for a recording is in the box, and it's portable. It's like taking your studio with you. Indeed, in most cases you don't even need an electrical source; many SIAB units have their own batteries. In addition to being self-contained and relatively easy to use, these systems include a keyboard and are very reliable.

But they do have shortcomings. First of all, most SIAB systems are very expensive. And with the rapid expansion of technology they can become outdated relatively fast. Furthermore, they don't have the flexibility that computer-based systems have.

A variation of the SIAB system is the "workstation." This is a unit that contains a sequencer, a keyboard, and usually a number of sounds. Again, it has the advantage of having the keyboard, sequencer, and sounds all in the same unit. Connections are still required for external units, but it's relatively easy to make high-quality, professional recordings using these units. But again, they can be expensive. (The terminology in relation to "workstations" and "SIAB" systems has become increasingly blurred in recent years, and the terms are sometimes used interchangeably.)

The third system is the one now used by most amateurs and many professional musicians, and it is definitely the cheapest. It is the computer-based, or software sequencing, system. One of the major expenses for this type of system is the computer, but most people already own a computer, so it's usually not a problem. Both PC and Mac computers work well with this type of system. The other major expense is a MIDI-equipped keyboard; these keyboards can range in price from a few hundred dollars to many thousands of dollars. Other requirements for a computer-based system are MIDI recording software, a MIDI-computer interface unit, perhaps some external sound modules, and a good sound card for the computer with a relatively large storage capacity.

A Primer on MIDI

Since MIDI plays such a large role in modern electronic music, it's a good idea to look at it in a little more detail. As we saw earlier, MIDI is a procedure for interfacing various types of instruments and musical devices. It is a digital system, so it uses digital data in the form of 0's and 1's (or "on" and "off"). The idea was first proposed by Dave Smith of American Sequential Circuits in 1981 and quickly generated a lot of interest. Today all music recordings are made using MIDI devices, but MIDI actually does much more than just record: it also controls many different hardware units. In particular, it allows synthesizers, computers, controllers, and samplers to exchange data and to control one another. We can, in fact, define MIDI as an extensive set of "musical commands" that can be used by electronic instruments to control one another. The key communication is a *MIDI message*. It consists of several data bytes similar to those in your computer. A group of these bytes is a message, which is nothing more than a series of numbers (in binary code) that we call a MIDI data file. MIDI data files are very small compared with normal audio files and are more convenient to use, as they take up much less storage space. The MIDI data file is transmitted in only one direction, and because of this the MIDI interface on any MIDI instrument must have two different MIDI connections: MIDI IN and MIDI OUT. In many interfaces there is also a MIDI THRU connection, which allows several MIDI devices to be "daisy-chained."

The MIDI data stream is usually generated in a MIDI controller, which can be a keyboard or a MIDI sequencer. Playing the keyboard generates a stream of MIDI data which is transmitted out through the MIDI OUT connections. From here it can go to several different places, but most commonly it will go to a MIDI sound module of some sort. This module will receive the message through its MIDI IN connections. The output from the keyboard can also go to a MIDI sequencer, where it is taken in through MIDI IN connections. The MIDI OUT connections on the sequencer are likely to be connected to one or more sound modules.

Let's look now at what MIDI commands do. As it turns out, there are hundreds of different commands or messages. If we are dealing with a keyboard and press a key—say, G—the first command will be a series of numbers telling an instrument to make a sound. We call it "note on." Since a piano has 88 keys, there have to be at least that many messages in order to specify any note. In practice, MIDI instruments can have up to 128 different notes. In the same way, there also has to be a "note off" command to stop the sound. When you press a key, the note sounded has a certain volume and other characteristics such as "aftertouch." These are also described by MIDI messages.

If these messages are passed to a MIDI sequencer, they can be stored and edited. Later, they can be replayed by passing the edited data from the sequencer via the MIDI OUT connections to a sound module. This sound module responds by playing back the same series of notes that you played on the keyboard.

This is only a brief overview of what MIDI is, and what it is capable of. We will look at it in much more detail in the next chapter.

Microphones

If you are planning on recording vocal music or an instrument that cannot be connected directly into a MIDI system, you will need one or more microphones. Microphones are, of course, also needed in live performances. You have three types of microphones to choose from: condenser, dynamic, and ribbon microphones.

These three types differ slightly in their response to sound waves, specifically, in regard to dynamic range, frequency response, and directionality. Dynamic range is the range of sound levels over which the microphone produces a useable electric signal that is acoustically proportional to the amplitude of the sound. Some mikes, for example, have a dynamic range of 0 to 100 dB, while others might have a range from, say, 60 dB up to 140 dB or more, so you have to be careful in your selection. The frequency response indicates how strong an electrical signal is produced for a given sound pressure at various frequencies. The ideal, of course, is a flat response over a relatively large range. Finally, directionality is a measure of how the microphone picks up sounds from various directions; the pattern is usually referred to as a polarity pattern.

Of the three types of microphones, dynamic microphones tend to accentuate the middle of the frequency spectrum. Condenser mikes tend to have a rounded shape to their frequency response, and ribbon mikes have a gradual reduction at high frequencies and smear sounds together slightly at low frequencies.

Condenser Microphones

As the name suggests, the main component of a condenser microphone is a condenser. Condensers (also called capacitors), which are used extensively in the electrical and electronics industry, consist of two plates with opposite charges on them. Between the plates is an electrical field, the magnitude of which depends on the amount of charge on the plates. In a condenser mike the front plate is the diaphragm; it is made of metal, metal covered with plastic, or sometimes Mylar, and is suspended in front of another metal plate called the backplate (fig. 108). A small voltage is applied to the diaphragm and backplate, producing an electric field in the space between them. When you sing or speak into the diaphragm, it vibrates; these vibrations change the distance between the plates and therefore the elec-

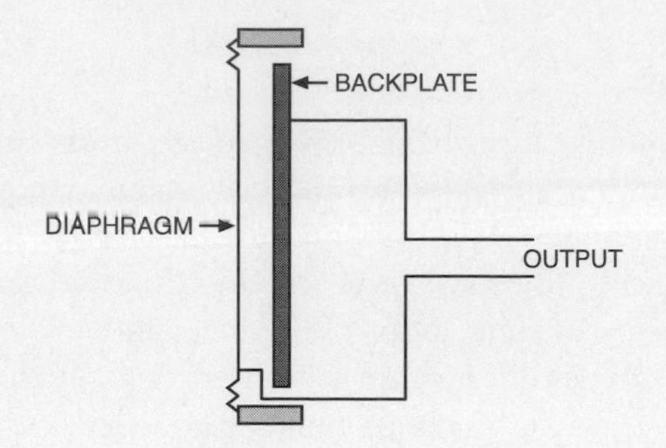

Fig. 108. A condenser microphone.

tric field between them. This variation produces an electrical signal that is sent to an amplifier.

Condenser microphones come in two main varieties: small diaphragm and large diaphragm. Small-diaphragm mikes have excellent high-frequency response and are therefore good for recording stringed instruments and acoustic guitars. Large-diaphragm mikes are used mainly for voice; they produce a warm, fuller sound with an excellent middle- and high-frequency response.

Dynamic Microphones

Dynamic microphones use magnetic fields rather than condensers (and electric fields) to convert sound to electrical energy. In particular, they use the same principle that is used in electric motors, namely, that a current-carrying wire in a magnetic field experiences a force whose strength is proportional to both the current and the field strength (and has a direction perpendicular to both). Alternately, this can be thought of as the interaction of two magnetic fields.

In the dynamic microphone, depicted in figure 109, the diaphragm is made of plastic or Mylar and is located in front of a coil of wire called the *voice coil* that is suspended between two magnets. When you speak or sing into the mike, the diaphragm moves, and in turn, this moves the voice coil between the magnets. The voice coil has a small current passing through it, and it changes when the voice coil moves within its surrounding magnetic field. The result is an electrical signal proportional to the amplitude of the sound of the voice.

Although dynamic mikes are not desirable for some types of recordings, they are particularly durable (they can stand up to a lot of

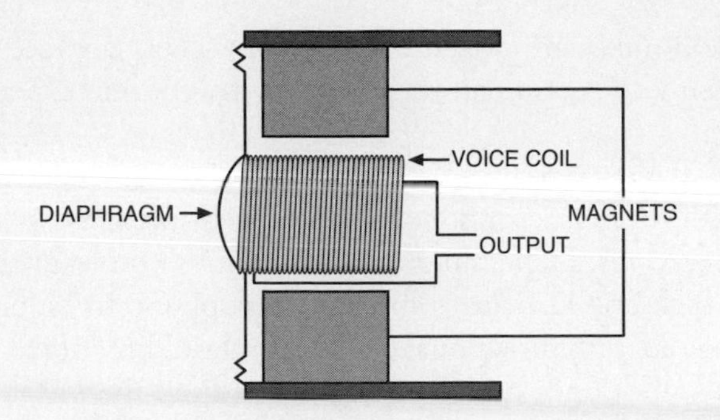

Fig. 109. A dynamic microphone.

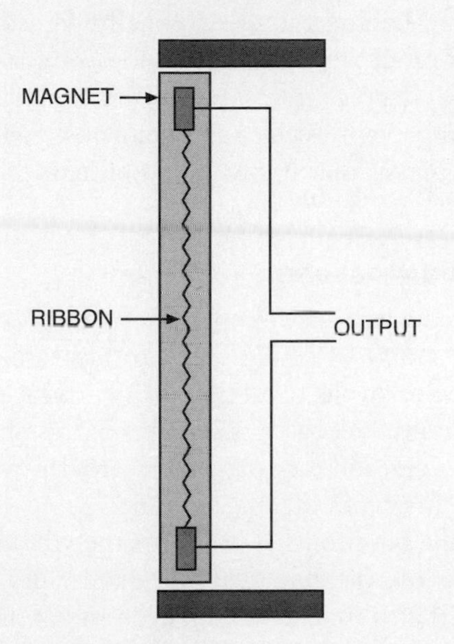

Fig. 110. A ribbon microphone.

abuse) and are therefore popular for live performances. They are also considerably cheaper than condenser mikes.

Ribbon Microphones

Ribbon microphones work on the same principle as dynamic microphones. The major difference is that instead of a voice coil, an aluminum ribbon is used (fig. 110). The diaphragm is in front of the ribbon. Ribbon mikes are prized by many musicians because they give a "silky," smooth sound, but they are very fragile and expensive.

Polarity Patterns

Each of the three microphones is capable of picking up sounds in specific directions. Depending on what sounds you want the microphone to pick up, you may want the microphone to be omnidirectional, cardioid, or bidirectional (fig. 111). Omnidirectional mikes are able to pick up sounds from 360 degrees around the mike. They are particularly useful when recording a large group of musicians or an orchestra. Cardioid mikes pick up sounds only from directly in front of them. They are useful when you want to control noise coming in from the side and behind the mike. Finally, figure eight mikes are bidirectional; in other words, they pick up sounds from two directions (usually 180 degrees from one another). They are particularly useful in recording two musicians simultaneously. Most microphones produce a weak signal that has to be amplified; this is done with a preamplifier.

Monitors (Loudspeakers)

Closely associated with microphones are monitors, or loudspeakers. Microphones respond to sound and produce an electrical signal; monitors respond to an electrical signal and produce sound. In short, monitors translate the electrical signal back into physical vibrations that create the sound, and they produce nearly the same vibrations as those in the diaphragm in the microphone.

The unit in the monitor that produces the vibrations is called the *driver.* It consists of a flexible *cone* (the diaphragm) made of paper, metal, or plastic that is attached to the wide end of the *suspension.* The suspension, in turn, is attached to the driver frame, called the *basket.* The narrow end of the cone is attached to a voice coil (similar to the one in a microphone). This coil is held in place by a *spider,* or ring of flexible material. These components are illustrated in figure 112.

Earlier, I explained how a voice coil works in relation to the microphone. I think it's helpful, however, to show that it can be explained in a completely equivalent but seemingly different way, and that's what I'll do here. The voice coil is actually an electromagnet; in other words, it is wire coiled around a piece of iron. When current is passed through the coil, the coil acts like a magnet. In the monitor

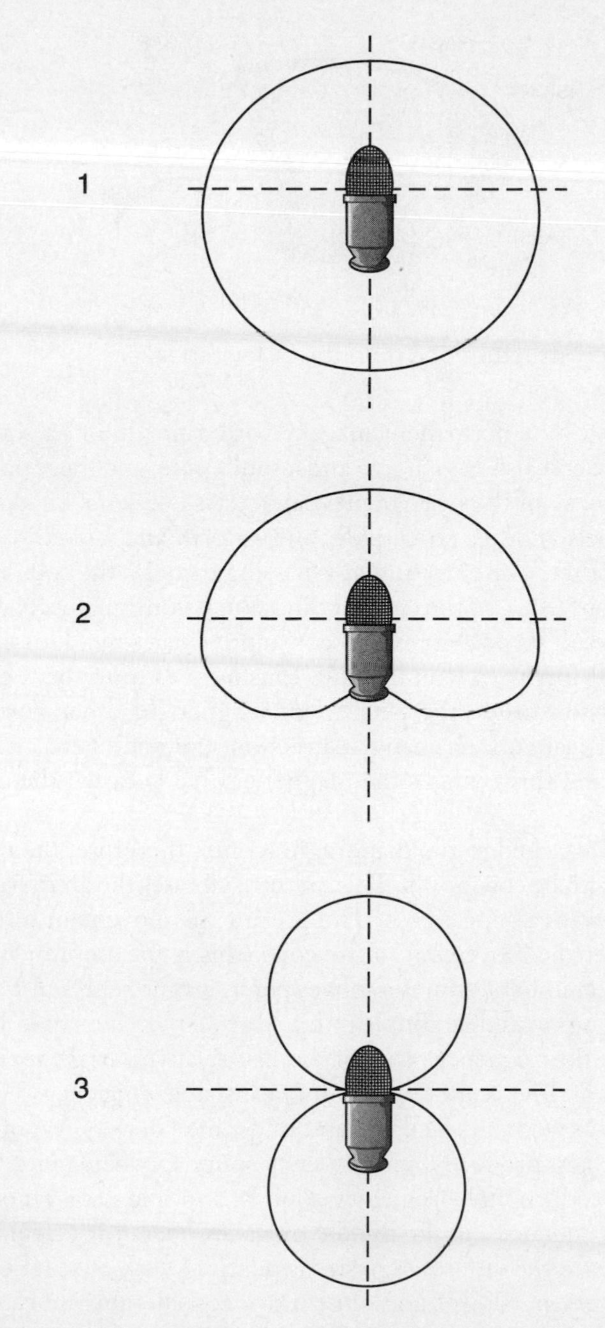

Fig. 111. Polarity patterns for a microphone: *top,* omnidirectional; *center,* cardioid; *bottom,* figure eight.

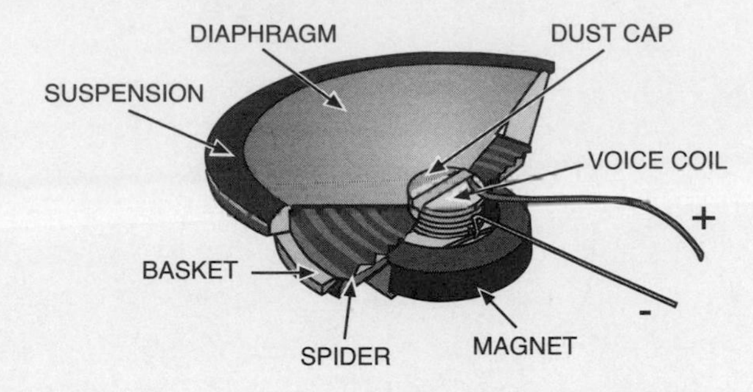

Fig. 112. Detailed structure of a monitor, or loudspeaker.

it is surrounded by a permanent magnet with a north and a south pole. The voice coil also has a north and a south pole, and these poles interact with those of the permanent magnet. As everyone knows, a north pole repels another north pole, and a north and a south pole attract one another. The current that is passing through the coil is alternating current (AC), so the current direction is continually changing, and as a result its polarity is also constantly changing. This creates a back-and-forth force on the coil, causing it to move back and forth like a piston inside the permanent magnet. In other words, when the electrical current changes direction, the coil's field direction reverses, and this changes the magnetic force between the coil and the magnet.

The coil is attached to the diaphragm (cone); therefore, the diaphragm moves as the coil moves. This, in turn, vibrates the air in front of the coil, creating sound waves. The greater the movement of the coil, the greater the movement of the cone (this is the amplitude of the wave). The number of times it moves per unit time represents the frequency. Things would be simple if this were all there was to it, but it isn't. Because the frequency range of sound is relatively large, several different types of drivers are required for reasonable efficiency. They are referred to as woofers, tweeters, and mid-range drivers. Woofers are the largest and produce low-frequency sounds; tweeters are the smallest and produce high-frequency sounds; and mid-range monitors produce frequencies in the middle of the frequency spectrum.

The reason for the difference in size is relatively easy to see. Low-frequency drivers are slower and must move considerably more air, so it makes sense that they are larger. High-frequency drivers have to move fast, so it is logical that they are smaller.

With two or possibly three drivers in a given unit, the signal has to be divided between them according to its frequency. The network for doing this is called the *crossover network* and consists of capacitors and inductors. There are two types of crossovers, active and passive. Passive crossovers don't need an external power source; they are actuated by the signal passing through. Active crossovers need an amplifier circuit in each driver. Passive types are the more common.

The drivers and crossovers are housed in a single unit called the monitor enclosure. It might seem that the overall structure of the enclosure would be of little importance, but it is actually very important. One of the major problems in loudspeakers is that the diaphragm is moving back and forth, so it's actually producing sound waves in both the forward and backward directions (in front of and behind the cone). These sound waves can interfere with one another. There are several ways of dealing with this in the configuration of the enclosure. One is called the *sealed enclosure;* in this case the enclosure is completely sealed so that the front sound waves travel out into the room, but those traveling backward are stopped in the box. In some designs the back wave is redirected so that it comes out in the forward direction and supplements the forward wave. Monitors with this type of enclosure are referred to as *bass reflex* monitors.

Dynamic monitors such as those I've discussed above are the most common, but other types exist. One type is referred to as the *electrostatic monitor.* In this case a large, thin conduction panel is used as a diaphragm. It is suspended between two stationary conduction panels that are charged. The panels create an electric field similar to that in the condenser microphone. The audio signal is passed through the suspended panel, and as it varies it moves relative to the back panel. Another type is called a *planar magnetic* monitor, which uses a long metal ribbon suspended between two magnetic panels, similar to the ribbon microphone.

Monitors of whatever type fall into one of two categories, based on their use: nearfield and farfield. The nearfield monitors sound best at a distance of three to four feet and are used extensively in studios; the farfield monitors are used at distances of more than four feet.

Recording the Sound: CDs and DVDs

For years LPs—long-playing vinyl records—were the standard devices for storing music. The sound was recorded on a spiral groove beginning at the outer edge and winding its way to the center. If you looked at the groove closely, it appeared to be wavy, or more exactly, it looked like a series of peaks and valleys at an angle of 45° to the surface. LPs were eventually replaced by magnetic tapes, which are still used in cassettes. Regions of varying magnetization are created on these tapes by magnetic fields from the alternating electric currents in the recording device. In other words, the individual atoms on the surface of the tape are oriented differently depending on the magnetic field that is applied to them.

Almost all music is now recorded on compact disks (CDs); most music stores now carry only CDs and cassette tapes (and DVDs). Music is recorded onto a CD in digital format; in other words, it is stored as a series of digits that allows the music to be recreated. Sampling is done at 44,100 times per second, and the information obtained is stored on a spiral track on a small disk. In this case, unlike LPs, the beginning of the track is not at the outer edge; it starts at the center and spirals to the outer edge. On the track is a series of small bumps and flat areas. The message is read using a small laser beam that is aimed at the track as the disk spins (usually at 200 to 500 revolutions per minute). When the laser beam is on a flat region, it is reflected back perfectly, but when it strikes a bump the light is dispersed. The result is a series of bright and dark regions. The reflected light is directed into a sensor that reads the bright regions as "on" and the dark regions as "off." Each of these on and off regions is a "bit" of digital information and is stored as a 1 (on) or 0 (off). The result is therefore something like 01001100110. Sixteen bits constitute a "word," and in the case of music CDs there are 65,536 word possibilities. The waveform of the sound is encoded using these words; the process is called pulse code modulation, or PCM. As the disk spins, 44,100 "samples" are collected from the right and left stereo channels each second. They are sent to a digital-to-analog converter (DAC), which takes the digits and uses them to produce a musical waveform. The signal is then sent to amplifiers and finally to loudspeakers, or monitors.

The year 1977 saw the release of a new type of disk that stores about 20 times as much information on the same disk size (about 5 inches). Called the DVD, it revolutionized the movie industry; movies could now be put on small disks—and of course, almost all of the movies ever produced can now be obtained on DVD.

14

Making a MIDI Recording

MIDI has now been around for over 20 years and has become central to recording for both amateurs and professionals. We had a brief introduction to it in the last chapter and will look at it in much more detail in this chapter. The feature we usually associate with MIDI is recording, but there is actually much more to it. Another important feature is triggering other MIDI devices, and it's also important in the storage and transmission of information about sound.

In its most basic form a MIDI system needs four things:

1. a sound generator,
2. a MIDI controller,
3. a sequencer, and
4. a MIDI interface.

The sound *generator*, as the name implies, is a device for generating sound. Several different devices fall into this category, including keyboards, drum machines, samplers, and sound modules. I will talk about each of them in considerable detail later in the chapter. The *controller* is what controls the MIDI instructions; it is, in essence, any MIDI device that controls other MIDI devices. It can be anything from a keyboard to a sequencer. The MIDI *sequencer* is central to the system, and since its main function is recording, it is sometimes referred to simply as the *recorder.* In practice, it can be either a hardware unit or software. Software sequencers are being used more and more in today's recording industry, so we will concentrate on them. Finally, a MIDI *interface* is required so that a computer can "talk" to the other MIDI devices in the setup.

Getting Connected

Turning to the details, let's look now at exactly what is needed for a MIDI system and how it is assembled. This depends to a large degree on the type of system you are planning to set up. Since I'm going to be talking mostly about software systems, I'll begin by listing the basic needs of such a system. They are

- a computer,
- a MIDI-equipped keyboard,
- MIDI sequencer software,
- a MIDI interface unit, and
- appropriate MIDI cables.

There are many possible setups, depending on how elaborate you want your system to be. Eventually you will no doubt want external audio, but I'll leave that to later. Figure 113 shows a schematic of a typical system.

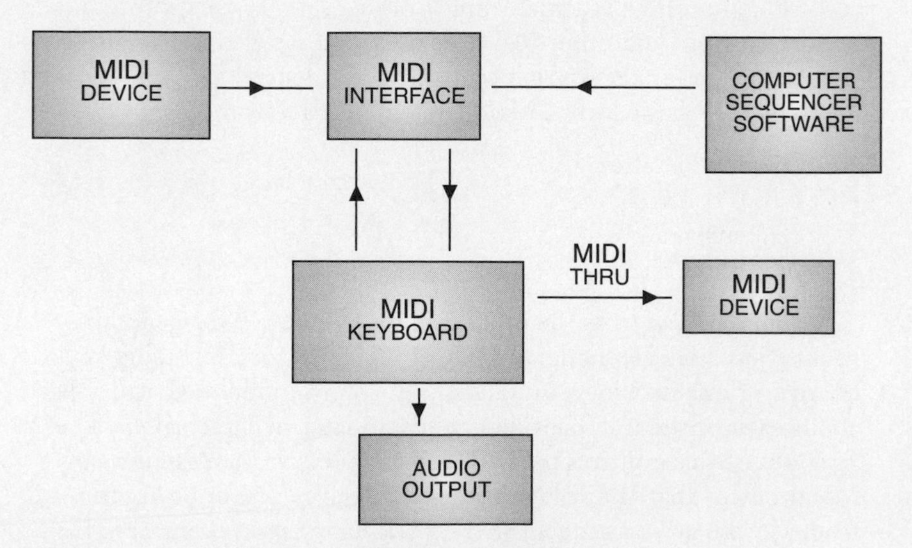

Fig. 113. Schematic of a typical MIDI system.

Computers are not usually equipped with MIDI IN and MIDI OUT connections; thus, you have to use either the computer's USB connection, or in some cases, the parallel or series connection. This requires a small box referred to as the MIDI interface. These boxes have MIDI IN and MIDI OUT connections and a USB connection for the computer and come in a variety of types and styles. The major difference between the various types of interface boxes is in the number of MIDI IN and OUT connections. The cheaper ones have only two, but boxes with four or even eight IN and OUT connections are also available. If you are using several external MIDI devices, you will need one with several connections.

Some interface boxes also have a MIDI THRU connection. If your keyboard doesn't have a MIDI THRU terminal, you may want one on your MIDI interface. Several devices can be "daisy-chained" using this terminal; in other words, they can be plugged into one another in a chain.

The cables used to connect MIDI devices have five pins (male) or five plugs (female). It's important to make sure all your MIDI devices are connected properly. A single cable carries information for 16 separate channels. For the various connections to work properly, the transmitter (the device sending the MIDI signal) and the receiver (the device receiving it) must be set to the same channel. As in the case of your TV set, you have to tune the system to the proper channel if you want so see or hear what's on that channel.

Figure 113 shows a couple of "MIDI devices." What type of devices are they? Many of them may be within your software if you're using a computer-based system, and if so, they will be stored somewhere in your computer until they are needed. Others, however, may be external hardware units. Drum machines are still used extensively as external units; they usually contain hundreds of different drum rhythms. In some cases these are synthesized sounds, but in others the sounds may come from samplers, where real drum sounds are used.

Samplers, whether external units or software, are used extensively in MIDI systems. They usually contain libraries of hundreds of different sounds and are now widely used in the recording industry. We will discuss them in considerable detail later. *Sound modules* of various types can also be used. These modules are usually stripped-down versions of a synthesizer or drum machine in that they have to be triggered externally.

One of the first things you have to do after you get everything connected is to *synchronize* all the units. This can involve several steps. First of all, you have to decide which units will send MIDI commands and which will receive them. The sender is referred to as the "master," and the receiver as the "slave." If, for example, you have two keyboards, you have to specify which one is the master and which is the slave. This is generally done using a dialing box within the software in your device. You may also have to synchronize your sequencer with various external units such as sound modules. In most cases your manual will tell you how to do this. Finally, if you are doing any audio recording, you will also have to synchronize the sequencer with the audio; again, the best place to find out about this is in your device manual.

Software for a Computer-Based Setup

One of the major parts of a computer-based setup is the software sequencer, sometimes referred to as the *production* software. At one time this component consisted of only a sequencer, but nowadays it usually contains much more, including numerous samples and other sound tracks of various types.

Software-based production systems are becoming increasingly common, mainly because of the cost, but they also have several advantages. First of all, they are limited only by the memory (or power) and speed of your computer, and modern computers are optimized for both these factors. In addition, computers' large screens are easier on the eyes than the small screens in many studio-in-a-box (SIAB) units and other types of units, and this makes editing much easier. As we'll see later, integrating audio into the system, synchronization, and mixing are all easier using computer-based systems. One of the greatest advantages, however, is that the software industry is advancing rapidly, and new, better, more user-friendly systems are appearing on the market every year.

I can't begin to cover all the software sequencers on the market, but they are generally similar to one another. Some are just more powerful than others, and as with most other things, you get what you pay for. Some sequencers are made only for Mac computers and some only for PCs, but in most cases there are editions for both Mac and PC. One of the best known software sequencers is SONAR, which is put out by Cakewalk. It is updated every year or so (as are most other systems). The latest version includes 64-bit audio recording and has a large number of instruments, effects, and so on. It also has a large number of tracks for both MIDI and audio.

Another popular software sequencer is Cubase; it, along with Nuendo, is put out by Steinberg. Both have large numbers of MIDI and audio channels and numerous virtual instruments and powerful mixing units. Sony manufactures two of the better-known software synthesizers, Sound Forge and Acid Pro, which have been particularly popular over the past few years. Acid Pro has "loop-based" software. A few others worth mentioning are Logic Pro, Peak Pro, and Live 5.

The system that is best for you depends on what you want to record, how elaborate a system you want, and what you can afford. Prices can range from a little over a hundred dollars for a minimal system to over a thousand dollars for a more elaborate one.

Making a MIDI Recording

I'll begin by giving an overview of MIDI recording and provide some details later. As we saw in the last section, there are many software processors, and they all differ slightly. I will give a general description of how to use them. You may find that the process differs slightly with the system you use (if you are using one), but bear with me: most of the main functions will be the same.

The first thing you do is load your software into your computer and make sure it is working properly. When you open the application, you will find several toolbar icons at the top of the main screen similar to those in most word processors. They control the recording, editing, and playing back of the MIDI sequences. It's a good idea at this point to give your song a working title. Then create a *track* for recording your song; you can call it "Track 1." To initiate the sequencer you may have to go to a new window, usually referred to as the *transport* window. Three of the buttons you will see here are Record, Play, and Stop. They act like the record, play, and stop buttons on any tape recorder: recording begins when you hit the record button and stops when you hit the stop button. To play back the recording, you use the play button.

When you have finished recording you will see a horizontal bar in Track 1 on the screen, as shown in figure 114. It will contain some vertical lines corresponding to notes that have been played. You can

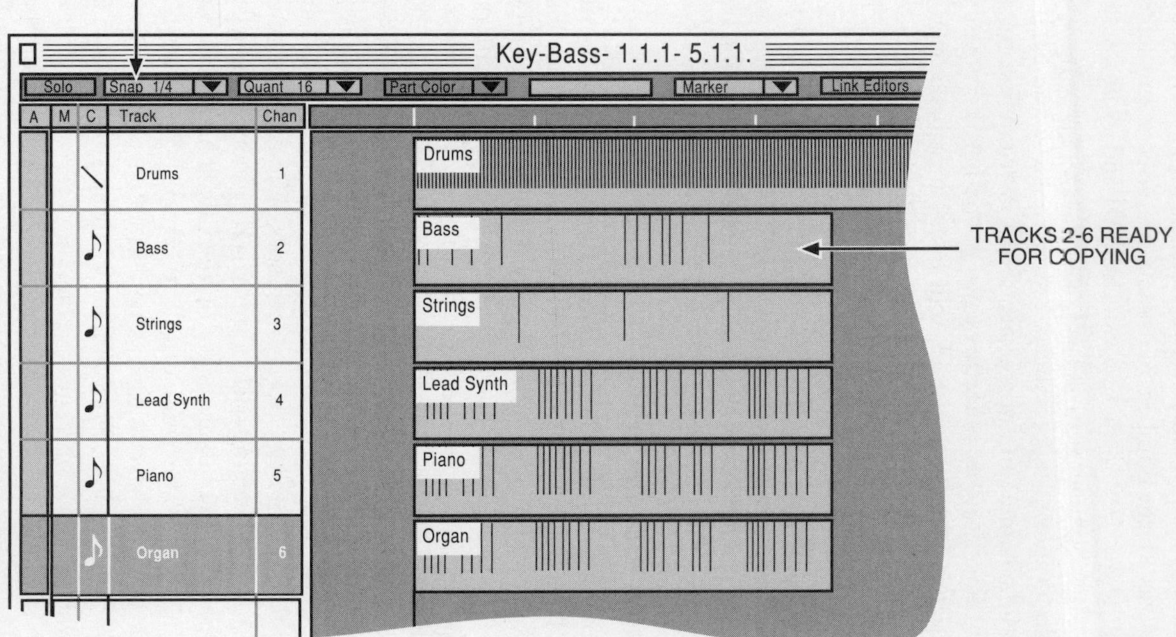

Fig. 114. The main window of a software recording application.

hear what you have played by clicking twice on the stop button (this may differ in some systems) and then hitting the play button.

In my instructions above, I'm assuming that you began playing in the usual way the moment you hit the record button; this is referred to as *real-time* recording. If you are unsure of yourself, however, you can do this differently: you can import the notes one at a time. It's a tedious process, but it can be done, and it's referred to as *step-time* recording.

One of the major differences between this process and the older recording methods (e.g., tape recording) is that if you make a mistake, you don't have to start over. You have the ability to edit the performance; if there is a bad note, or bad timing, you can change it. In fact, the process is just as simple as making corrections on your word processor. To edit, you double click on the horizontal line; this opens what is called the "piano roll window," and in this window you will see a keyboard to the left and a number of horizontal lines in the center of the screen, as illustrated in figure 115. If, for example, you have recorded four bars it will show all four bars and they will appear to be small segments of horizontal lines. Each note that is struck is, in fact, shown by a horizontal bar, and the length of the note is represented by the length of the bar. You can also see which note was struck by looking at the keyboard to the left.

You will also see a background grid in the piano roll window, so it's like a graph with the horizontal axis being time and the vertical one being frequency (pitch). The grid helps you determine the type of note (eighth, quarter, half, and so on) that has been played and is invaluable in editing.

To make changes in this window you merely highlight the bar you want to edit and then change it. You can shorten or lengthen the bar, or move it.

If you prefer working with the musical notes (the score), there is another window that can also be used for editing: the *score* window. When you call it up, you see the notes that have been played written out as a musical score (fig. 116). This window is usually the one preferred by musicians. In the same way as in the piano roll window, you can alter the notes by changing their position or deleting them. One of the easiest ways of making changes is by highlighting the note you are interested in, then dragging it to a new position or deleting it. In some sequencers you can also print out the score.

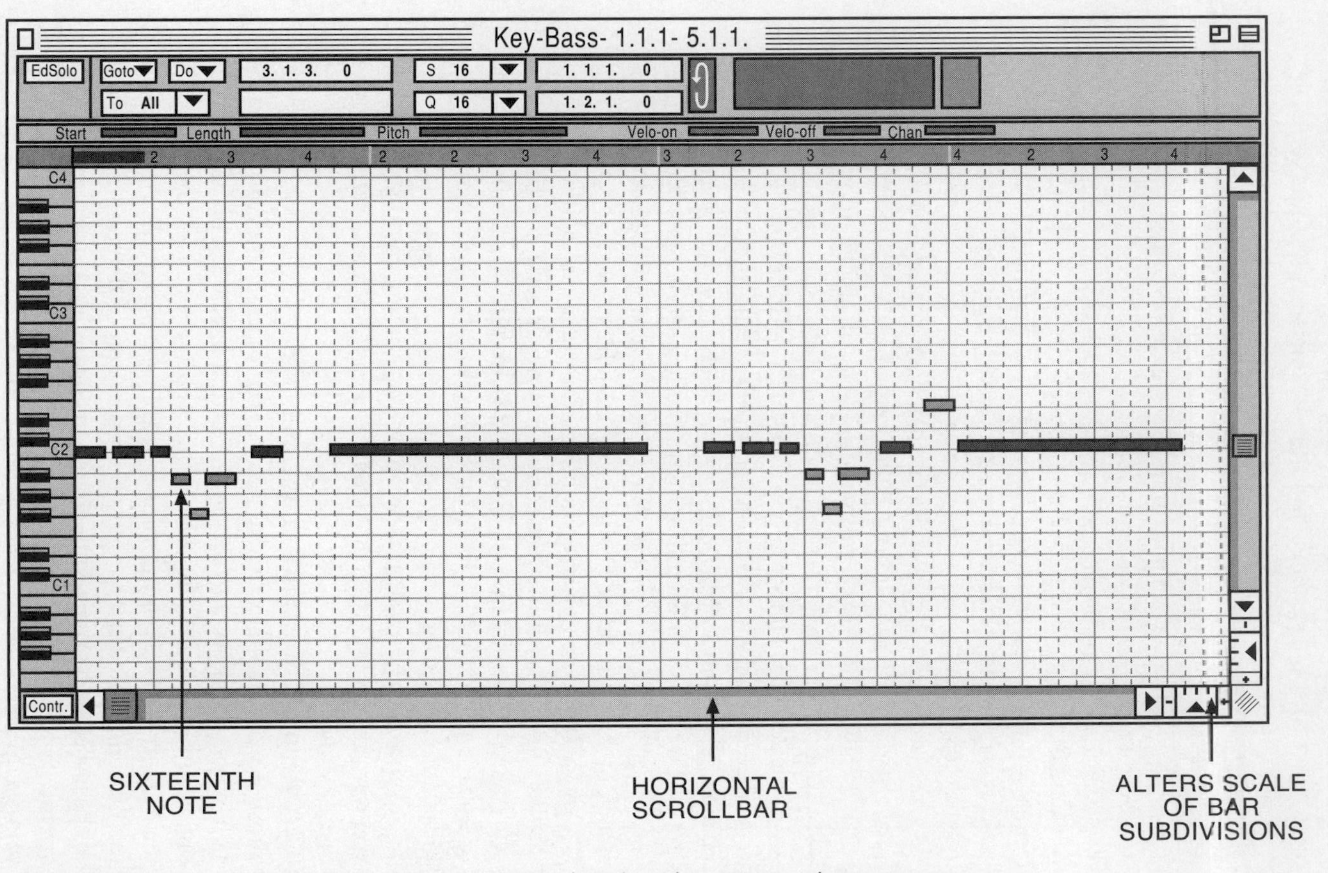

Fig. 115. The piano roll window of a sequencer showing the notes as bars.

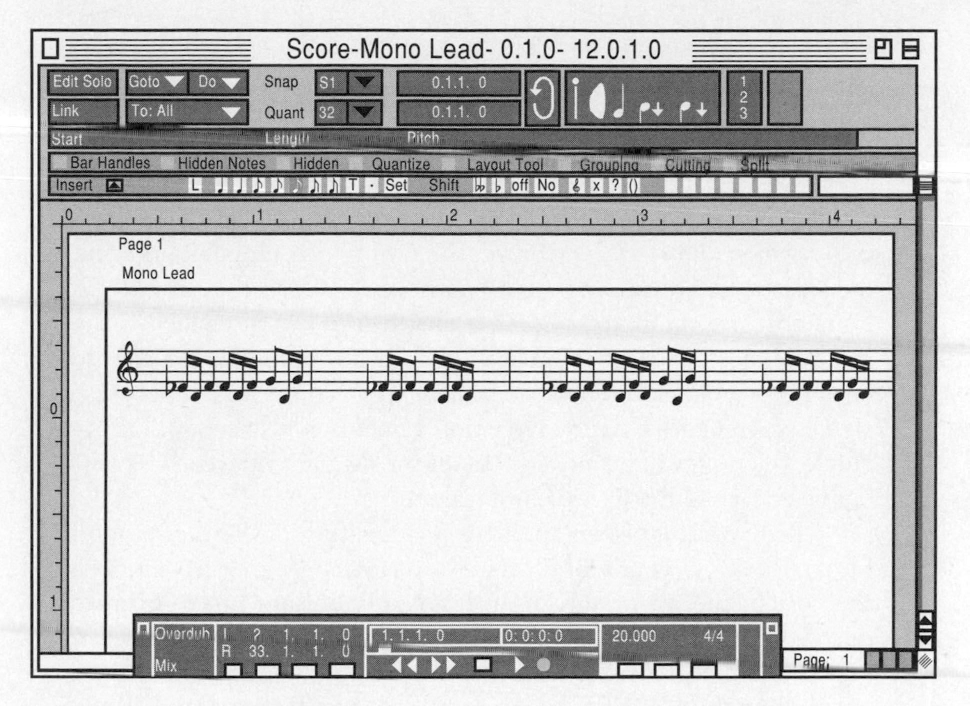

Fig. 116. The score window.

There's actually a third window that can be used for editing, the *events* window. It gives you a list of the MIDI events that make up the track. The notes played are not seen here, only a large number of vertical columns, and they can be changed as in the case of the two previous windows. This window is not used extensively by musicians for editing, but computer "geeks" may prefer it because of its list of the MIDI "events" in the track.

Timing and Transposition

One of the problems many musicians (I'll exempt professionals here) have is timing. In fact, the position of the notes in either the piano roll or the score window will quickly show you how far your timing is off (and it may embarrass you if you haven't played with others very often). Consider the piano roll window; using the grid on the screen it's easy to see where the notes are falling. You can see each bar, and each is divided into four beats; furthermore, each of these beats is divided further into 4, so you can see eighths and sixteenths. It's therefore easy to make the needed corrections: just move the bars to where they should be.

Timing can also be corrected using the score window. In this case it is merely a matter of moving the notes to where they should be. Unfortunately, this can be a tedious process. But you can get around it using one of the corrective features on the menu called "quantization," which allows you to correct the timing of a range of notes or, indeed, of the entire song—but you have to be careful using it. To initiate quantization you choose a quantization value, which is the subdivision to which the note is moved. If, for example, you choose 8, all notes will be moved to the nearest eighth note. You can also assign a percentage of the quantization value; if you select 50, for example, the note will be moved 50% of the distance between its present position and the quantization value.

When you're finished applying quantization, it's always a good idea to look over your score carefully. Quantization may have moved the note in the wrong direction. It's also important not to overuse quantization. No one plays a score perfectly, and if it's too perfect, it may sound mechanical or what we sometimes call "clinical."

Another feature of sequencers that is sometimes useful is the transposition feature. If you played the song in the key of C, for example, it allows you to transpose it to another key. Songs do, in fact, usually sound better in some keys than others. Musicians usually have to transpose by moving all the notes in the score a certain number of half tones up or down from the given note. In years past there were pianos that could transpose music mechanically using a small knob; Irving Berlin is said to have had one. He wrote all his songs in the key of E-flat, then used the changer to listen to them in other keys before deciding which key he wanted to publish them in. On a computer this is done quite simply.

The final task after you have edited your track and are satisfied with it, is of course saving it. This is done in the same way that you save anything on a computer; you merely hit the "save" button. You will have no problem playing back what you saved on the same sequencer, but if you plan to transfer the file to a different sequencer, you need to take certain steps in advance. To make sure it can be played properly on a different sequencer, you have to save it as an SMF file (standard MIDI file). In addition, to make sure that the sound that you produce on your machine sounds the same on all machines, you have to make sure that both pieces of equipment bear the logo "GM" (for General MIDI).

We now have one track filled, so it's time to look at other tracks. To make full use of your sequencer you'll have to fill several tracks: that is, of course, the idea of the sequencer. One track is usually reserved for rhythms, and you'll probably want to program drums on this track. They can come from several sources: a drum machine, a sampler, or drum sounds stored in another device. A bass line is also important in relation to rhythm; it may be placed on another track. Other tracks can be used for sounds such as strings, organ, and brass. When you have several tracks filled and edited to your satisfaction, it's time to move on to the next step.

Let's assume you have recorded eight bars of each of the tracks. What's next? One thing you may want to do is duplicate them all into the next eight bars. This is easily done using the Copy and Paste commands, as with a word processor.

Looking again at the other channels, I mentioned that each of them will have sounds of instruments, but where do these sounds come from? They can, in fact, come from several places. Samplers, drum machines, external sound modules, and softsynths are the usual sources, so let's turn to them.

Samplers and Softsynths (Virtual Instruments)

Samplers play a large role in the recording industry today, so let's begin with them. As we have already seen, a sampler is an electronic musical instrument similar to a synthesizer in that it produces the sound of many different instruments, but it does not produce these sounds from "scratch." A sample of an instrumental sound is recorded, then changed and modified in several ways. Because it comes from "real" instruments, it is frequently much more realistic than synthesized sounds. Samplers can be either hardware or software; we'll begin with the hardware units. The earliest samplers (for example, the Kurzweil 250 and Korg M-1) had keyboards associated with them, but later ones were usually controlled by an external keyboard. Various keys on the keyboard accessed various sounds, or samples.

Samplers can play back many different kinds of recorded audio, but they do much more than merely playing back recorded sounds. Most have editing capabilities that allow the programmer to apply a wide range of effects. The first sampler was the Mellotron; it came out in 1976 and was followed in 1979 by the first polyphonic sampler, the Fairlight CMI, which was made in Australia. These instruments could play a sound from a tape when a particular key on the keyboard was struck; needless to say, they were limited (and costly). By the late 1980s, however, several low-cost Japanese samplers had come on the market.

Over the next few years samplers became much more sophisticated. Most had extensive editing capabilities and were able to record and manipulate segments of tracks for looping and reprocessing. Looping is the process of re-recording the same part of a sound several times; it is used extensively in dance music. Drum sounds, in particular, are frequently looped. Segments can be copied, cut, and pasted as with a word processor.

Samplers now contain many of the filtering features that were common in early analog synthesizers. Filters of various types, along with frequency oscillators and envelope generators, can be applied to samples or even groups of samples to change their sound. Another important function of samplers is stretching time. Samplers can alter the frequency (pitch) of a sound by speeding up or slowing down the speed at which it is played. As we saw earlier, if you double the speed, the frequency moves up by an octave. There is, however, a problem with this. If, for example, the sample contains a background rhythm, such as that from drums, speeding up will also speed up the drums and throw things off. This is taken care of by a process called "time stretching," in which tiny pieces of data are removed (or added) at equal spacing along the sample to allow for the change in timing.

Most samplers have a screen in which the editing is done, but there is a serious problem with such screens: they are very small. You can get around this by transferring the video to a computer screen, which is much larger.

In recent years hardware samplers have begun to be replaced by software samplers. These applications are fully functional samplers but need no hardware; they use the processing power of the computer, and they can do everything a hardware sampler can do. They do, however, require a lot of computer memory, so a high-grade computer is needed.

Closely associated with software samplers are software synthesizers, also known as *softsynths* or virtual instruments. It is, in fact, hard in many cases to tell the difference between software samplers and softsynths. In theory, softsynths are software versions of synthesizers and therefore produce artificial musical sounds, whereas software samplers produce real musical sounds, but in reality most softsynths are now heavily sample-based; in other words, they use samples extensively. Some of the sample-based softsynths, in fact, come with huge libraries of sounds. Not only do they produce the sounds of traditional musical instruments, but they can also produce the sounds of popular early synthesizers that are no longer manufactured. Two of the sample-based softsynths are the MiniMoog and the Yamaha DX-7. When they emulate synthesizers, they are frequently referred to as *emulators*.

Mixing and Mixers

We opened several tracks in a sequencer in our earlier discussion. Our next project is to bring the tracks together to make a single recording that contains all of them. Before we begin, though, I should mention that you can actually record one track on top of another in a process called *overdubbing*. Simply, it is adding a track to one that has already been recorded. The process is straightforward: you start with the usual procedure for recording, but use a track on which you've already recorded rather than a new one.

The process of bringing the tracks together is called *mixing*, and it is done with a mixer. There are many different hardware mixers, but we'll be concerned with software mixers, which come with your production software. The program that appears on your computer screen looks like a hardware mixer, with a large number of knobs, dials, sliders, and so on (fig. 117). All the tracks that you recorded earlier will be here.

The mixer allows you to do many different things. For example, you can change the dynamics (volume) of each of the channels according to your taste. You can also bring in other channels (instruments) at various points in the recording, or take them out. Another important control is the equalizer (EQ), which allows you to adjust the tone on each of the channels. There are controls for fading in and out and panning. As the name suggests, the fade control allows you to let the signal fade away. Panning controls how much of the signal goes to the right and left stereo speakers.

Equalizing is one of the most important functions of the mixer. It allows you to adjust the frequencies of the instruments in the various

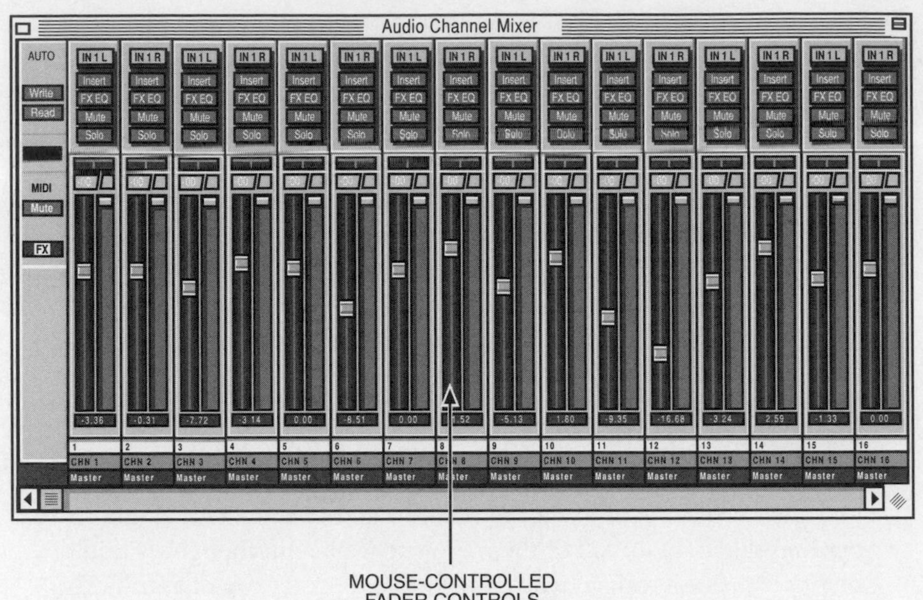

Fig. 117. The mixer.

channels so that they do not conflict with one another. Adjusting the equalizer is similar to adjusting the bass and treble tones on a hi-fi system. The effect here, however, is a little more dramatic. Your goal in using the equalizer is to reduce the frequencies that clutter the track and enhance those that make an instrument sound better. In effect, you have to give each instrument with a similar frequency range a little space. There are two basic types of equalizers: parametric and shelf. The shelf equalizer targets a range of frequencies above and below a particular frequency. It is used to cut off frequencies at the top and bottom of the frequency spectrum. The parametric equalizer, on the other hand, allows you to target any frequency range. You can select a particular frequency and a small range around it and boost or decrease it. It is a particularly useful tool.

The dynamics control is also of considerable value. The loudness of instruments within any arrangement varies. This control allows you to make adjustments according to your taste. The best way to do this is by dropping the levels of background instruments, then bringing them in and adjusting their loudness. It is important to listen for tonal quality and the effect each instrument has on the overall sound until you are satisfied.

Sound Effects

Another important aspect of mixing is adding special effects such as reverberation, delay, and chorus to the overall mix. Without special sound effects your song is likely to sound flat. The main sound effects are reverberation, delay, phasing, flanging, chorus, ADT, echo, and distortion. Special effects can be produced by hardware units, but we'll consider only the software that produces them. Like the software programs for mixing, special effects programs look the same on the computer screen as the hardware units. The effects are usually called "plug-in" effects and are implemented by tiny programs that are called up and run by the recording software.

Of all the effects, *reverberation* (usually referred to as "reverb") is the most important. It is the natural acoustic effect that occurs when sounds bounce off surroundings before they fade away; reverberations are most easily heard in an empty building. It is reverb that gives "life" to your sound, so it is particularly desirable to make sure it is added properly.

Another important effect is *delay* (which actually comes in several forms). It is created when sound waves bounce around a room, and like reverb it is a natural effect. An effect called *phasing* occurs when a signal is played from two different sources at the same time. If the sources were side by side, the peaks and troughs of the wave would line up, but when they come from separated sources there is cancellation that creates a "sweeping" sound. The effect is created when there is a delay between the two signal of 7 to 12 millisecs. A particular type of phasing, where the delay is 12 to 20 millisecs, is called flanging. In this case the sweep is more pronounced.

Another type of delayed effect is called *chorus*. As the name suggests, it gives the effect of fullness to a voice or an instrument. Chorus is produced by modulating a number of repeats; in other words, one or more copies is made of the original track that are slightly out of tune with it; then they are played back together.

Echo occurs when a signal is sufficiently delayed that a separate sound can be heard. It is used extensively with vocals and guitars and is produced by lengthening the delay. Other effects are *pitch shift* and *distortion.* Distortion is used mainly by guitarists, but occasionally by keyboard players. It is created by playing the signal so loud that distortion occurs and then re-recording it at reduced volume.

Recording Audio

So far we have been talking about recording MIDI from various MIDI instruments. You can, for example, record MIDI from a MIDI keyboard or a sound module. But what if you want to record voice or an acoustic piano? You obviously need something to get the sound into your system, and one of the main things you will need is a microphone. Consider what happens when someone sings into a microphone. The microphone converts the changes in air pressure into voltage changes, but in the MIDI system these voltage changes have to be changed into digital data. So you need a converter, and it is contained in the *sound card* in your computer. Sound cards are small, integrated circuit boards that are slotted into your computer's PCI slot.

A sound card of some sort will come with your computer but it may not be powerful enough for what you want. Sound cards, in fact, vary considerably, from basic devices to those used by professionals that contain multiple audio outputs and other special features. (The Macintosh computer doesn't use a sound card; the equipment for integrating audio is built into it.) If you need a more advanced sound card than the one that is in your computer, you will have to take off the outer cover and insert it, but this is usually a relatively simple task.

Once your sound card is installed, you can record audio by connecting the input cables from your microphone or amplifier to the input connections on your sound card. (Two other ways of doing so are through an interface that is connected to your USB port or through a port called the FireWire port, but these methods are slower than the direct connection to the sound card, so I will assume that you are using a sound card connection.)

We will begin by plugging the audio source from the microphone into the desired channel on the sound card; then we will select a computer track on which to record. Your sequencer program should have both MIDI tracks and audio tracks available. These two types of tracks are different but are side by side in your computer. What are some of the differences between them? First of all, digital audio cannot be displayed as notation, as MIDI can. This means there is no score window or piano roll window—what you see on the screen are "waves" of sound—but the process of recording is basically the same as for MIDI. After checking sound levels and so on, you press the Record button and play (or sing). When you come to the end, you press the Stop button.

Finally, I should say a few things about placing the microphones when you make an audio recording. Where you place them relative to the sound source makes a tremendous difference. It is, in fact, a skill that is achieved only after considerable experience. There are three common placements for mikes: close miking, distant miking, and ambient miking.

In close miking the microphone is usually only inches away from the sound source. This placement is frequently used to avoid picking up background sounds from the room. In distant miking the microphone is three to four feet away from the sound source; this placement allows you to record some of the sounds of the room (reverb and echo) along with the voice or instrument. In ambient miking the microphone is placed even farther away so that most of the room effects are captured.

In most cases you will record to two microphones for a stereo effect. There are several different techniques for this; one is referred to as X-Y pairs, where the two microphones are right next to one another. In other cases they are widely separated. You will have to experiment to find out what separation is best for you.

So far we have said little about the recording room—namely, the studio—and its acoustics. This will be covered in the next chapter.

15

The Acoustics of Concert Halls and Studios

In 1895 Harvard University opened a new lecture auditorium that was considered to be an architectural masterpiece, but it was soon discovered to have extremely poor acoustics. Very little was known about acoustics at the time, and the university wasn't sure how to fix the problem. So Harvard asked Wallace Sabine, a young professor in the physics department, if he could find out what the problem was. Sabine had no background in acoustics, but he, along with several assistants, began a series of experiments and soon discovered that the time for a sound to die to inaudibility was particularly important; he referred to it as the *reverberation time.* Over a period of a couple of years he gained considerable insight into the acoustics of halls and was able to significantly improve the acoustics of the Harvard auditorium.

Sabine's accomplishment drew the attention of the people designing a new hall for the Boston Symphony, and he was brought in as a consultant to make sure the acoustics were as good as possible. He had learned a lot in working with the Harvard auditorium, and he applied it to the Boston hall: he measured the reverberation time carefully, made allowances for changes when the hall was full of people, and was sure that his recommendations would produce excellent acoustics. The hall opened in 1900, and everyone was curious to see how good the acoustics were. To Sabine's dismay, he was severely criticized by several music critics. Although he had considered the presence of the audience, he had slightly underestimated the effect, and the reverberation time was a little different from what he had assumed it would be. He was so devastated by the criticism that he never worked in acoustics again and never mentioned the hall. The irony is that 50 years later the Boston Symphony Hall was regarded as one of the best acoustical halls in the world and literally everything he did was correct. He is now considered to be the "father" of building acoustics, and the unit of sound absorption, the sabin, is named after him.

The Basic Principles of Acoustics

Let's begin our look at acoustics by considering a large room and a source of sound. We'll assume we are somewhere near the center of the room. The acoustical energy from the source moves at the speed of sound (340 m/sec, or 1,116 ft/sec) outward from the source in straight lines. It spreads out in a sphere around the source, which means that the further out it is from the source the weaker it is. In fact, acoustical energy obeys what is known as an *inverse square law*, which says that when the distance increases by 2 the energy weakens by a factor of 4. This tells us that the energy has weakened considerably by the time it reaches you, assuming you are somewhere near the center of the room.

Sounds waves from the source itself will be the first to reach you, usually in a tiny fraction of a second. At the same time, however, sound waves are moving out in other directions and are striking the ceiling and walls. As a result they are reflected, and some of the singly reflected waves soon strike you, then waves that have been reflected twice strike you, and so on. The reflected waves will have less energy than the ones that come to you directly. Nevertheless, in a relatively short period of time, the entire room will be bathed in waves coming from all directions. The most obvious result of all these reflections is that the source will sound louder (as compared to a similar source in the open air, where there is no reflection).

When a sound wave strikes a surface some of the energy is reflected, some is absorbed, and a small amount may be transmitted through the wall (we'll ignore this transmitted energy for now, since it is small). One of the things that is of particular interest is how much of the acoustical energy is reflected and how much is absorbed. It's well known that hard surfaces such as concrete, marble, and stone reflect most of the energy, while soft surfaces such as curtains, carpeting, chipboard, and acoustic tile absorb most of it. The types of surfaces have to be taken into account when we are considering the reflection and absorption of the energy of the wave as it moves around the room.

Also important is the size of the room. The larger the room, the farther sound has to travel before it strikes a surface and the farther the reflected waves have to travel to the observer, and we have to take this into consideration.

So let's consider an experiment. Assume we are in a relatively large room at some distance from the sound source. We will assume that the source is a percussive sound—short and sharp—and we'll make a plot of the intensity of the waves that strike us and the time that it takes for them to strike us. The result is what we see in figure 118. In the diagram we see the waves directly from the source; the waves that have been reflected once, twice, and so on; a large buildup that occurs from all the reflections; and finally, decay of the sound.

Let's compare this to what happens if, instead of a short, sharp sound, we have a steady source of sound. In this case there will be a buildup, as shown in figure 119, and initially it will be fairly irregular. Finally, however, a steady state will be reached in which the sound level is constant. In this case the energy losses by absorption within the room are balanced by the energy being released by the source, and as long as everything stays the same, this will continue indefinitely. What happens, however, if we suddenly turn off the source? The sound does not suddenly go to zero, as you might expect; it decays as shown in figure 120.

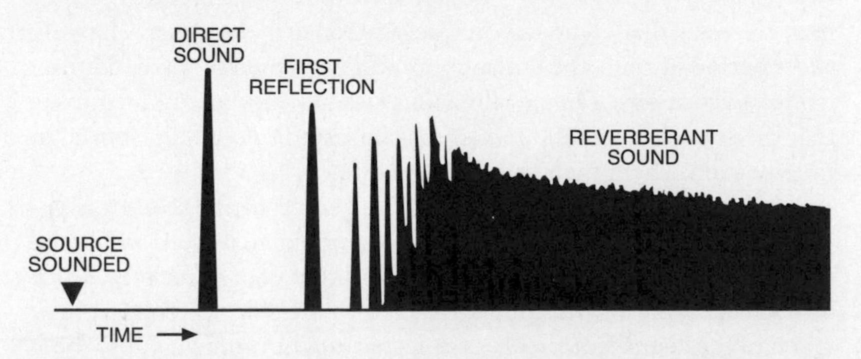

Fig. 118. A plot of intensity versus time for a short, sharp sound.

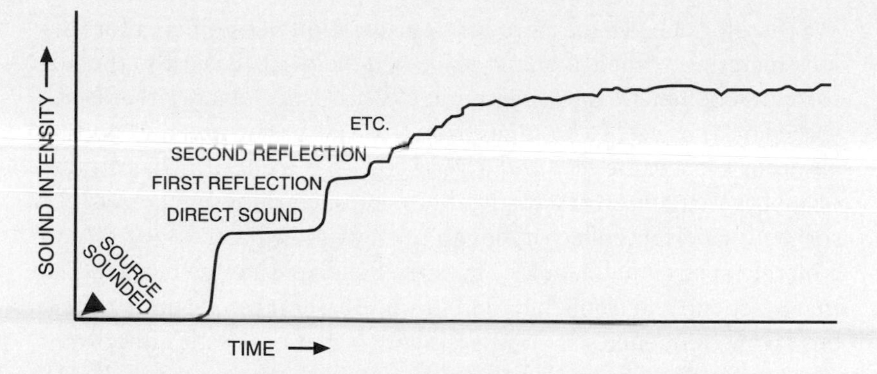

Fig. 119. A plot of intensity versus time for a steady source of sound.

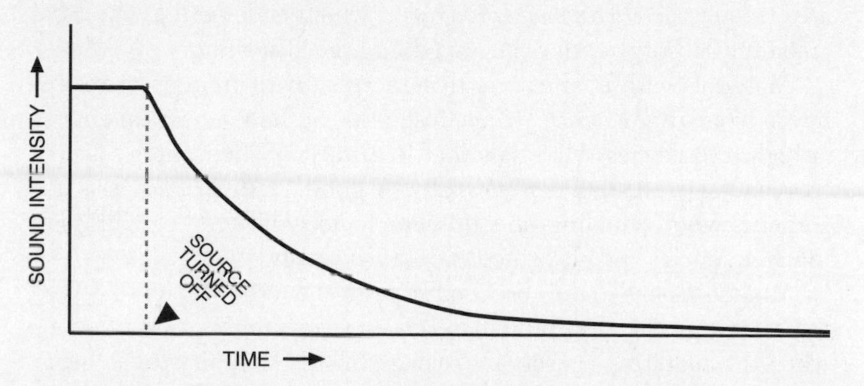

Fig. 120. The decay of sound when the source is turned off.

The time that our source takes to get to inaudibility depends on several factors, but for a particular hall it will always be the same regardless of what intensity we start with. This is the time that Sabine called "reverberation time." We now define it a little differently than he did. It is formally defined as the time required for sound to drop to one-millionth of its initially intensity, or to drop in loudness by 60 dB. This time plays a central role in the acoustics of all buildings.

Reverberation Time

As we saw above, reverberation time depends on the absorption of the walls, ceiling, and floor of the room, and of course, the amount of absorption that takes place depends on the area that is absorbing.

We therefore need a measure of the absorption per unit area for various materials, called the *absorption coefficient* (α). Basically, the absorption coefficient is the percentage of acoustical energy that is absorbed by a surface, so it ranges from 0 to 1 (it has no units). A perfect absorber has a value of 1 and a poor absorber (reflector) has a value of approximately zero. Actually, there's no such thing as a perfect absorber or a perfect reflector, but an open window has a value of 0 and a marble surface has a value of about .01. Experiments have shown that α depends on frequency, and in some cases it can be quite different at low frequencies as compared with high ones. So it's important when discussing absorption that you specify the frequency. Table 13 shows the absorption coefficients of various materials at five different frequencies. When we multiply the absorption coefficient by the area we get the *effective absorbing area*, which is defined as the total amount of absorption for the area doing the absorbing.

Materials with coefficients that increase with frequency are referred to as *treble absorbers*. They lose most of their acoustical energy at high frequencies. Materials that lose most of their energy at low frequencies are called *bass absorbers*. A good example of a bass absorber is wood paneling on studs; the panels will tend to vibrate at low frequencies and absorb considerable energy.

What happens to the absorbed acoustical energy? It is transferred to heat within the absorbing material, a process that is most effective when the material has small air spaces in it. Such air spaces occur most commonly when a material is made up of small fibers. The en-

Table 13. Absorption coefficients of common building materials at various sound wave frequencies

	Frequency (Hz)				
Material	125	250	500	1,000	2,000
Marble	.01	.01	.01	.01	.02
Smooth concrete	.01	.01	.01	.02	.02
Brick	.03	.03	.03	.04	.05
Painted concrete	.10	.05	.06	.07	.09
Plaster on concrete	.10	.10	.08	.05	.05
Wood floor	.15	.11	.10	.07	.06
Acoustic tile (ceiling)	.80	.90	.90	.95	.90
Heavy curtains	.15	.35	.55	.75	.70
Plywood on studs	.30	.20	.15	.10	.09
Carpet over concrete	.08	.25	.60	.70	.72

ergy is actually absorbed by the air that is trapped between the fibers. Even though the sound waves are transformed into heat, the actual change in temperature of the material is small. Absorption is also increased if the surface of the material is covered with small holes or cavities. The energy is easily trapped in the cavities. Acoustical tiles combine these two features: they are made of fibrous material and have small holes in their surfaces.

In summary, then, both the coefficient of absorption and the total absorbing area are important in relation to the total absorption, and as we saw, their product is referred to as the effective absorbing area (A). Also important, however, is the rate at which the sound energy strikes the various surfaces before it is reflected and absorbed, and this depends on the volume of the room. If the room is large, the energy will be spread over a large volume and the energy density will not be as great when it strikes a surface. Sabine brought these two effects together into a formula that gives us the reverberation time, T_R. It is given by

$$T_R = .16\,V/A,$$

where V is the volume and A is the total absorption of all surfaces.

In practice there is another contribution that can be important, and it is the absorption of the air within the room. It is given by the volume of the room (V) times the absorption coefficient for air, which is quite small (on the order of .003). Air absorption is important only if the volume of the room is very large, and it is also greater at higher frequencies. If it is needed, a contribution $\alpha_{air}V$ has to be added to A.

Finally, in calculating A you usually have to add together the contributions from many sources, including each of the walls, the ceiling, and the floor.

Examples of Reverberation: A Concert Hall and a Small Room or Studio

Let's turn now to a sample calculation of reverberation time. I'll give two examples: a concert hall and a smaller room. Consider the concert hall first; assume it has the dimensions 30 m × 40 m × 10 m high, so that its total volume, V, is 12,000 m^3. We'll assume that one of the 30 × 10 walls has a curtain over it, that the other wall is plywood on studs, that the two 30 × 40 walls and the ceiling are plaster over concrete, and finally, that the floor is carpeted. We'll also assume a sound wave frequency of 1,000 Hz. Multiplying the area of each of these surfaces by the absorption coefficient for the material (provided in table 13) gives us the effective absorbing area:

Wall, 30 × 10 (curtain)	300 × .75 = 225 m^2
Wall, 30 × 10 (plywood)	300 × .10 = 30 m^2
Two walls, 30 × 40 (plaster over concrete)	2,400 × .05 = 120 m^2
Floor, 30 × 40 (carpeted)	1,200 × .70 = 840 m^2
Ceiling, 30 × 40 (plaster over concrete)	1,200 × .05 = 60 m^2
Total absorbing area	1,275 m^2

So, the total absorbing area in this concert hall is 1,275 m^2; these units are also referred to as "sabins." Using the formula already provided for reverberation time,

$$T_R = .16\ V/A,$$

we can calculate the reverberation time, T_R, for this space as

$$T_R = .16\ (12{,}000)/1{,}275 = 1.5 \text{ sec.}$$

As we will see later, a reverberation time of 1.5 seconds is quite reasonable.

Now let's turn to the smaller room; it might, for example, represent a studio. We'll assume it is 4 m × 6 m × 3 m high, so that its total volume is 72 m^3. Assume further that the walls are plaster on concrete, the floor is carpeted, and the ceiling has acoustic tile on it. Again taking the absorption coefficients for each of these materials from table 13 and multiplying by the area of each surface, we arrive at the following absorption areas:

Two walls, 4 × 4 (plaster on concrete)	32 × .05 = 1.6 m^2
Two walls, 6 × 6 (plaster on concrete)	72 × .05 = 3.6 m^2
Ceiling (acoustic tile)	24 × .10 = 2.4 m^2
Floor (carpeted)	24 × .07 = 1.7 m^2
Total absorbing area	9.3 m^2

We then calculate the reverberation time as

$$T_R = .16\ (72)/9.3 = 1.2 \text{ sec.}$$

We'll look at the significance of these numbers a little later, but for now let's go back to the concert hall and consider the effect of an audience on reverberation. The acoustics of a hall are quite different when the hall is full of people compared with when it is empty. This was, in fact, part of Sabine's problem in relation to the Boston Symphony Hall. People in the audience absorb sound energy at a relatively high rate. Experiments show that the average person has an absorbing area of approximately 0.5 m^2. (Again, this varies with frequency, from an absorbing area of about 0.35 m^2 at 125 Hz to 0.6 m^2 at 5,000 Hz; therefore, the frequency of the sound must be taken into consideration.) If there were, for example 1,000 people in our auditorium, they would add an effective absorbing area of $0.5 \times 1{,}000 = 500$ m^2. The reverberation time, T_R, would then be

$$2{,}119.2/1{,}775 = 1.19 \text{ sec},$$

which is considerably less than the 1.5 seconds when the hall is empty. Thus, the presence of the audience definitely affects the acoustics.

The best way around this problem is to have seats with an absorbing area approximately equal to that of a person, so if the seat is empty it will make little difference. And indeed, this is what is done. Unfortunately, it's difficult to match a person exactly; upholstered seats generally have an absorption coefficient of about 0.3.

Ideal Reverberation Times

So, what is the ideal reverberation time? As it turns out, there is no reverberation time that is ideal for all situations. The ideal time depends on several things, and in the case of music it depends on the type of music that is played in the hall. In general, it depends on the clarity needed; short reverberation times are best for piano music, and longer ones for organ and orchestral music. The shortest times are needed for speech because it has to be particularly clear to be understood. If the reverberation time is very short, we refer to the hall as "dead." Music played in a dead hall does not sound satisfactory, but speech may be okay.

A plot of ideal times for various types of music is given in figure 121. The reverberation times of some well-known concert halls at frequencies of 500–1,000 Hz are as follows:

Boston Symphony Hall	1.8 sec
Cleveland Severance Hall	1.7 sec
New York Carnegie Hall	1.7 sec
Philadelphia Academy of Music	1.4 sec

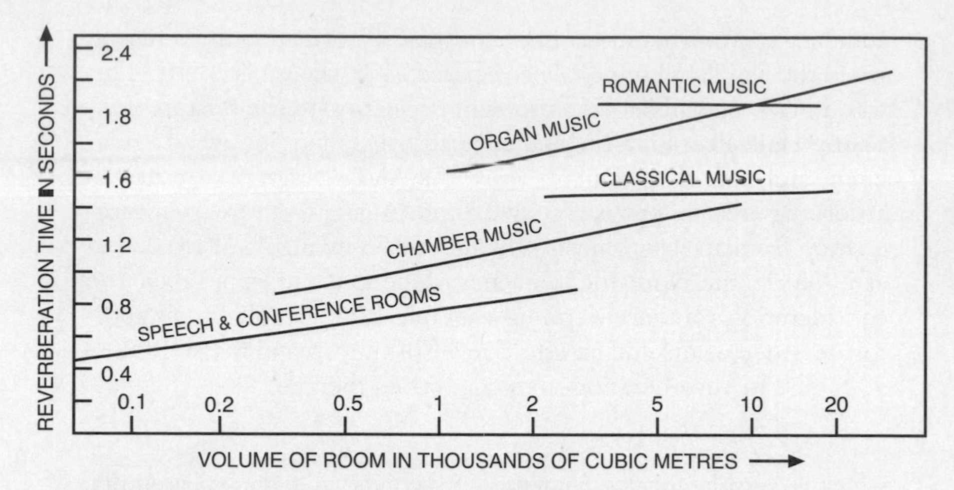

Fig. 121. Ideal reverberation times for various types of music.

The formula for reverberation time is only approximate and works best when the sound is uniform throughout the hall. It tends to give incorrect values for a hall with a highly irregular shape, a hall in which all the surfaces are highly absorbing, or a hall where the absorbing surfaces are all concentrated in one area.

Finally, the average reverberation time for a residential room is about 0.5 sec. Short reverberation times are preferred for radio and TV station newsrooms, usually on the order of 0.1 sec.

Other Criteria of Importance in Concert Hall Acoustics

While reverberation time is the most important criterion in relation to the acoustics of concert halls, several other factors are also important. The first of these is *intimacy*. Intimacy is related to the time separation between the direct sound and the earliest reflections, which is usually referred to as the *initial time-delay gap*. Halls are generally considered to be acoustically intimate if this time delay is less than 0.03 sec, which is the shortest time the human ear can distinguish.

Another important criterion is *warmth*. In this case we are concerned with the bass frequencies. A hall is said to be warm if the reverberation time of the lower frequencies (50–250 Hz) is considerably higher than that for the high frequencies (above 250 Hz).

A third factor is *brilliance*, and it depends on the prominence of high frequencies (above 2,000 Hz) in the hall. For brilliance, the hall must have a reverberation time for the higher frequencies that is greater than that for middle and low frequencies.

Another consideration is *fullness of tone*. This factor depends on the reverberation time and the ratio between the loudness of reverberation sound and that of the direct sound. For fullness of tone this ratio must be high. Strangely, for *clarity*, which is important in speech, this ratio must be small. So it's obvious that halls that have fullness of tone have poor clarity and vice versa.

Finally, a problem that sometimes occurs in large halls is *echo*. It usually comes from the back wall and can easily be reduced.

Correcting Acoustical Problems

The question that now comes to mind is, How do you improve the acoustics if they need improving? Since acoustics depend critically on the reverberation time, that should be the first thing you look at. As we saw earlier, the best values are around 1.7 sec; if reverberation time in the hall you are considering differs considerably from this, changes will have to be made. There are two cases that have to be considered: permanent changes needed because of acoustical flaws in the hall and temporary changes to accommodate fluctuating numbers of people in the audience.

Permanent acoustical changes can be made by changing the material of the interior surfaces—for example, by covering the ceiling with acoustic tiles. Temporary changes are made in several different ways. One is to use moveable acoustic curtains that have a high absorption coefficient. They can be rolled out when needed. Some concert halls are equipped with wood panels on the side walls that can be opened to expose highly absorptive materials. Suspending reflective panels called *clouds* is also frequently used to improve acoustics. In some cases they are suspended directly above the stage; in this case, they help the players hear one another more clearly. They are also of help in changing the initial time-delay gap, so they also help the intimacy. In fact, most of the other acoustical problems noted above, such as lack of intimacy, lack of warmth, and lack of brilliance, can easily be improved by adjusting the reverberation time.

The Shape of Halls and Acoustics

Another thing that has an effect on the acoustics is the shape of the interior of the concert hall. When the surfaces are flat, sound waves are reflected in the same way as a light beam reflected from a mirror. In other words, the angle of incidence of the beam is equal to the angle of reflection (see figure 122). When you are dealing with curved surfaces, however, sound waves are reflected in quite different ways. Consider a spherically concave surface, for example. In the case of light we know that parallel beams striking the surface will be focused to a point. (More exactly, only a parabolic curve will focus them exactly to a point, but a spherical surface focuses then approximately.) This means that if a sound wave strikes a concave spherical surface, that surface will tend to focus them. In general, this is something we want to avoid. But it is helpful in the case of outdoor band shells; the curved walls behind the band reflect the sound forward toward the audience. This increases the intensity of the sound in the forward direction, but it can cause problems. In the case of most band shells, the high frequencies are reflected much better than the low frequencies.

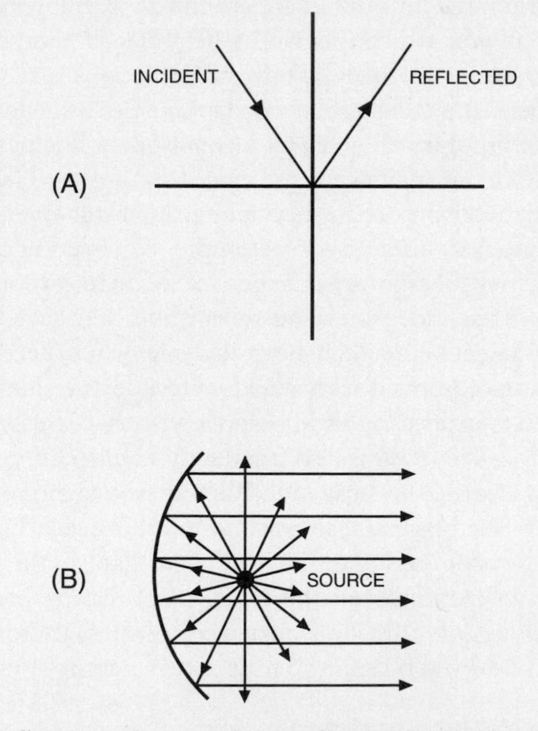

Fig. 122. Reflection of a sound wave from a flat surface (A) and a curved surface (B).

A number of well-known auditoriums have elliptical shapes, which can have a serious effect on their acoustics. Most of these auditoriums were built long before acoustics were well understood. The Mormon Tabernacle in Salt Lake City, for example, was built in 1867. Because of its elliptical cross section, it has two points called foci that appear to have amazing acoustics. If someone with a pin is posted at one of the foci and you are at the other, when the pin is dropped you can easily hear it hit the floor. Such designs are frequently referred to as *whispering galleries.* Despite its strange cross section, the Mormon Tabernacle actually has relatively good acoustics. Another well-known whispering gallery is St. Paul's Cathedral in London, which was built in 1668.

The Acoustics of Studios

In the preceding chapter we talked about how to set up a MIDI studio in terms of what equipment is needed and how to set up the equipment. But we said little about the studio itself and of its acoustics. The acoustical issues for a small studio are quite different from those of a large concert hall.

Let's begin by looking briefly at a professional sound studio. As we saw, reverberation times are important in recording music, and an ideal reverberation time for a large hall is about 1.7 sec. Because of this, most professional recording before the 1960s was done in large halls. But after techniques were developed for adding reverberation and echo (and other effects) into the recording during the mix, studios changed significantly. A typical professional studio now consists of several rooms. The vocalists and instrumentalists perform in a room referred to as the "studio." It is isolated from what is called the "control room," which houses the equipment for recording, routing, and manipulating the sound. Most studios also have "isolation booths" for particularly loud instruments such as drums and electric guitars. Other rooms are also sometimes used.

Let's turn now to a smaller studio, one that might be used by amateur musicians or even professionals who prefer to do their own recording. A small or home studio is likely to be a single room. In setting up a home studio, the first choice you have to make is the selection of the room. As we'll see below, the acoustics of this room are important, although reverberation does not play the central role that it does in concert halls. You should select a room in a basement, or away from street noise, that is as quiet as possible.

The next thing you must address is soundproofing. Not only do you want to isolate the room from outside sounds, but it's also a good idea to make sure you're not disturbing the neighbors too much. Even though the close-microphone techniques frequently used today don't pick up much background noise, soundproofing is still important when you are recording audio. When you're recording an electronic instrument such as a keyboard or guitar that is plugged in, it is not as critical.

To soundproof the room you need to seal the major sound leaks in it. It's a good idea to start with the ones around the doors and windows. Rubber weatherproofing strips or tape can be used for this. If the door is hollow, it should be replaced by a solid one, or a panel can be placed across it. Heavy curtains across windows or doors are usually of considerable help in reducing sound. Finally, things such as outlets and air ducts should also be covered.

Reverberation is not as important here, but it should be shorter than in concert halls. For music coming from loudspeakers, which you will have in your room, a reverberation time of around a second is usually adequate. For recording a speaker it can be as short as half a second.

One major problem in home studios is standing waves. They are created by parallel walls, and of course we have three pairs of parallel surfaces (two sets of walls, and the ceiling and floor). Acoustic waves are reflected back and forth from these parallel surfaces, and as they interact with one another, they may set up standing waves. We can, in fact, calculate the frequencies at which this is likely to occur. Using our formula $v = \lambda f$, along with the known speed of sound (340 m/sec or 1,130 ft/sec), we can write

$$f = 1{,}130/2L,$$

where f is the frequency affected, L is the distance between the parallel walls, and 1,130 is the speed of sound in ft/sec. Consider two walls 30 feet apart. We get

$$f = 1{,}130/2(30) = 28.2 \text{ Hz},$$

so it's obvious that the most serious problems will occur at low frequencies. One of the basic lengths in a room is the floor-to-ceiling distance; it is usually 8 feet, and the frequency affected in this case is 70 Hz. This may not seem to be much of a problem, but integral multiples of these frequencies can also set up standing waves. In the case of the 20-foot walls this means that frequencies of 56, 84, and 112 Hz (and so on) may also be problems.

There are two ways of getting rid of standing waves. One is by using *absorbers*, and the other is by using *diffusers*. We discussed absorbers in considerable detail earlier in this chapter. Obviously, if most of the energy is absorbed by the wall, standing waves will not be a problem. Absorbers are usually made of foam, with grooves or designs of some type on them. Fiberglass insulation also makes a good absorber. It's not a good idea, however, to fill the walls with absorbing material; doing so will make the room too "dead." Diffusers, on the other hand, diffuse or spread out the energy; not only do diffusers prevent standing waves, but they also help eliminate "dead spots" within the studio. Diffusers have pyramid shapes or lattices of various types on their surfaces that reflect the sound waves in many different directions.

Another problem related to standing waves is *phase interference*. If sound waves arrive at a particular point out of phase, certain frequencies will be reinforced, and others will be canceled or weakened. This sort of interference is usually countered by moving the speakers. Figure out where the major reflections are occurring and use it to stop the interference. This brings us, in fact, to where the speakers should be placed in the room. For optimum performance they should be about 8 feet apart and generally about the same distance from the listener. It's also important to have them an equal distance from the walls and not too close to a wall.

I've already indicated above that too many absorbers in the room can create problems. Absorbers generally absorb at middle and high frequencies much better than they do at low frequencies; if you use too many, you may cut out the upper frequencies and be left with reflections that are mainly in the low-frequency range. Most absorbers, in fact, work only down to 100 Hz, and carpet on a pad absorbs down to 250 Hz, so it is easy to see why low frequencies are left. *Bass traps* are designed to absorb these frequencies.

Also useful in some cases (particularly if the sound is dead) are reflectors. They are generally simpler than absorbers or diffusers. A sheet of plywood makes a good reflector.

So how do we "tune" our studio so that it has good acoustics and a flat frequency response across the spectrum? The best way is to use *acoustic panels* in your studio. They can be bought, or you can make them; they come in various sizes (a typical one might be 2 × 6 feet). They should be placed strategically around the room. Place two absorbers near or behind the speakers to help alleviate interference. Absorbers mounted at the opposite end of the room are also sometimes helpful. The ideal room has a mixture of absorbers, diffusers, and reflectors, so that no one surface is very "dead" or very "live." The only way you can get what you really want is by experimentation.

Epilogue

Well, we've come to the end of the book, so let's look back over what we have learned. I trust, of course, that you have learned something. First of all, we've seen that there is, indeed, an intricate connection between physics and music, and it goes deeper than you might have thought. From a simple point of view, music is sound, and sound is an important branch of physics, but the connection actually goes much further than that. I hope you have come to realize that as you read the book. If your primary interest is physics, you may have been surprised that music is tied to physics in so many different ways. If, on the other hand, you are a musician, with only a peripheral interest in physics, I'm hoping you found some insight into and appreciation of the fundamental role physics plays in music. I also have to include mathematics here, as we saw that the notes of scales and chords obey an important mathematical relationship.

We began by looking at the relationship between sound and music. In particular, we looked at many of the properties of sound such as loudness (intensity), frequency (pitch), and the fact that sound, and therefore music, is a wave and has wave properties. We also looked at the concepts of interference, reflection, and refraction of sound waves, all of which are important in the study of acoustics.

We saw how musical scales are set up and looked at several types of scales. Besides the usual diatonic scale, two that are of particular interest to musicians are the pentatonic and blues scales. Furthermore, we saw how notes are brought together into chords, discussed various types of chords and chord sequences, and learned how and why they are important in music.

As I'm sure you knew before you read the book, music comes in many different genres, ranging from rock and roll, pop, country, through jazz to classical. In chapter 7 I surveyed most of the main types and discussed briefly how they differ. I hope I didn't miss your favorite.

From there we went on to a survey of the various musical instruments, beginning with the piano. In each case I showed how the instrument worked and ended the chapter with a brief survey of some of the virtuosos on the instrument. Included in this survey were the piano, the stringed instruments (the violin, in particular), the brass instruments (including the trumpet and trombone), and the woodwinds (including the clarinet and saxophone). Finally I talked about the voice. It may not be an instrument in the usual sense, but it certainly plays a central role in music.

Modern music has become increasingly electronic in recent years, and the book would not have been complete without a discussion of electronic music. Electronic music has had a large impact on the music industry in recent years. Of particular interest in this regard is MIDI, which as we saw, is important not only to professional musicians but also to amateurs.

We finished up with a chapter on acoustics, which showed how we can measure the acoustics, not only of large concert halls, but also of small music studios.

So that's it. But wait! What about some of the other things that have influenced the music industry in the last few years? There are, indeed, several, and two of the most important are iPods and MP3 files (and of course I have to include the Internet here, since it is closely associated with both of the above). No, I didn't forget about them, and so we will briefly discuss both of them in the next two sections.

iPods

The iPod, whose familiar image you see in figure 123, is one of the major ways that people now listen to music. To give you some idea of how influential (and important) iPods have become, consider the

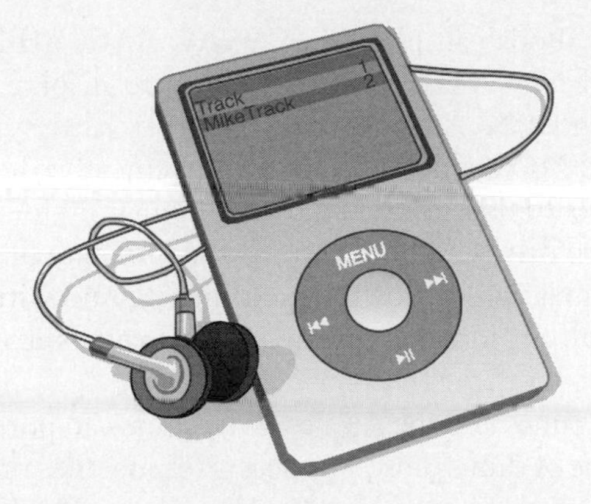

Fig. 123. An iPod.

following: as I write, 100 million iPods have been sold since their introduction in 2001. That's not a number to scoff at. More people may now be listening to music on iPods than on radios.

iPods were introduced by Apple Computer in October 2001. At first it may have seemed that they were just another digital audio player, but what surprised everyone was that they had a 5-GB (gigabyte) hard drive capable of holding over a thousand songs, yet they were so small they could easily fit in your hand. What made the device particularly attractive was that songs could be downloaded from the Internet via the "Apple Store" on an application referred to as iTunes. Thousands of songs were available by this means; in addition, songs could be downloaded to the iPod from CDs.

You may wonder where iPods got their strange name. One of the salesmen responsible for marketing the device was a science fiction fan and remembered a line from the movie *2001: A Space Odyssey*. It was, "Open the pod door, Hal," referring to one of the pods in the spaceship *Discovery*. Apple decided to tweak the word "pod" for its new device, and as the saying goes, "the rest is history."

Compared to the powerful iPods of today, the first ones were toys. They have now progressed through five generations (as of 2007) and have gone from a memory of 5 GB up to 160 GB. Furthermore, the latest ones are capable of video in addition to audio, and some can now be hooked into the Internet. And surprisingly (or perhaps not with the rapid advance of technology today), as they added more features and got more powerful, they got smaller and lighter. In addition, a touch-sensitive click wheel has now been introduced.

The latest iPods can play MP3, WAV, AAC, AIFF, and several other types of files. They have the ability to display JPEG, BMP, GIF, TIFF, and PNG image files, and they also support H.264, MPEG-4, and MOV video files. More specifically: the 80-GB iPod can store up to 20,000 songs, up to 100 hours of video, and 25,000 photos; its hard drive can hold various types of data files; and a large number of games can be downloaded into it. And with all this, it is less than a half an inch in thickness and weighs a whopping 5.5 ounces.

Through iTunes and the Apple Store, users can purchase 3.5 million songs; tens of thousands of podcasts (digital files on specific subjects, many of which are free); 3,000 music videos; 20,000 audio books; numerous video games; feature films; and even TV shows. Most of these specifics apply to the larger video iPods; the two major ones now on the market are the 30-GB and 80-GB models. They both have video screens, but smaller iPods without screens are also available. Two of them are the Nano (8 GB) and the Shuffle (1 GB).

Let's look briefly at the major parts of an iPod. (The specifics, of course, differ, depending on the model.) They are the following:

- *The hard drive*, which can store up to 80 GB in the larger models
- *The battery.* The battery is a rechargeable lithium ion battery. As the name suggests these batteries use lithium ions (charges). Lithium is the lightest metal and has a high electrochemical potential (the voltage difference within an electrochemical fluid) that is ideal for batteries. Lithium itself cannot be used because it is unstable.
- *The screen*, a 2.5-inch LCD screen. LCD screens are commonly used in watches and small computers. They use two sheets of polarizing material (material that exhibits different properties in different directions) with a liquid crystal solution between them. The passage of an electrical current through the liquid causes the crystals to align so that images can be formed.
- *The click wheel.* This wheel is a plastic touch-sensitive wheel with an embedded electrical grid and mechanical buttons.
- *The microprocessor*, a silicon chip that contains a CPU (central processing unit) that controls the system
- A *video chip* that controls the video
- An *audio chip* that controls the audio

One of the most amazing parts of the iPod is the touch-sensitive click wheel. It provides two ways to input commands: you can slide your finger around the wheel or you can press buttons located under the outer ring of the wheel and at the center of the wheel. There are four buttons under the wheel that activate the following functions: menu, back, forward, play/pause.

The click wheel's touch-sensitive function allows you to adjust the volume, move through various lists, fast-forward through songs, or fast-reverse, all by moving your finger around the wheel. But how does it work? There is a membrane with metallic channels embedded in the plastic cover of the wheel. And since the channels constitute a conducting grid, they are like a *graph*, and therefore each position on the wheel has an *address*. The sensing is done through the phenomenon called capacitance.

You're probably familiar with capacitors (also called condensers) in electrical systems; they store electrical charge. From a simple point of view they are two electrically conducting plates that are separated so that no charge can pass directly from one to the other; figure 124 provides a simple representation of this. Each has a wire connected to it (so it is part of an electrical circuit). Electrical charge flowing in the circuit builds up on one of the plates (let's say it is negative charge), and it attracts an electrical charge of the opposite type (positive) to the other plate. In an AC circuit the current is continually changing, so the effect of the changes is transmitted through the plate via the changes.

In the click wheel your finger is like one of the plates. When it touches the surface of the plastic, it forms a capacitor with the section of the grid directly below it. Current wants to flow to your finger to complete the circuit, but it is stopped by the plate (which is an in-

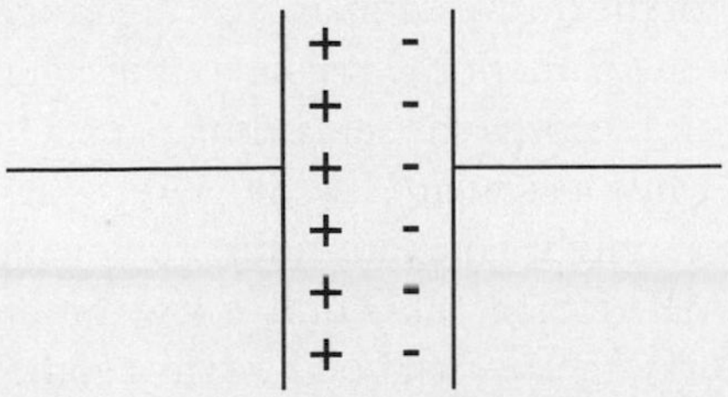

Fig. 124. A simple representation of a capacitor.

sulator). Because of this a small charge is built up at the point of the electrical grid just below your finger, and this tells the device where your finger is. Furthermore, a controller within the device measures the capacitance, or change in capacitance, as you move your finger, and sends a signal to the iPod microprocessor telling it what your finger is doing. When the microprocessor receives the message, it performs the action that you specified (e.g., increase the volume or create a list).

MP3 Files

MP3 is a buzz word you've no doubt heard many times in the past few years, and it is important in relation to the iPod because the iPod processes music in the MP3 format. To explain what this is, let's begin with a brief look at the various types of computer files. A music or audio file isn't much different from any other type of file on a computer. Basically, it's a digital data file that obeys a specific set of rules that defines how all the zeros and ones of the digital code are to be stored on the hard drive. The set of rules is referred to as the "format." We discussed the MIDI file format in the last chapter. In addition to MIDI files there are other audio files, including WAV files, AIFF files, and several others. They are in what is called a "noncompressed" audio format.

Now, suppose you have a song in the WAV or AIFF format, or on a CD, and you try to download it to (or from) the Internet. It would take an hour or more (probably much more) to do this, so if you wanted to download a dozen songs you'd obviously have a problem. Few people, in fact, would have the patience to wait it out. A way around this was discovered in 1979 by engineers in Germany and the United States about the same time. Their idea was to "compress" the music. What this compression does is remove sections of the sound track that are out of the normal human hearing range. In effect, compression discards the parts of the music that are generally inaudible (to most people) and records the remaining part. The resulting file is an MP3 file. So, while a regular CD (or WAV or AIFF file), in noncompressed format, might take four hours to download from the Internet, music in MP3 format takes only a few minutes.

The obvious drawback is that you have lost some of the "fidelity," or quality, of the recording, and the resulting music will not sound quite as good as it does coming from the original CD. But in most cases the difference is relatively small. You would notice the difference if you compared the MP3 version to a CD version played on a relatively good hi-fi set. But most people who download music from the Internet play it on systems that are incapable of exhibiting the extracted part, so the two sound about the same.

To change a recording to the MP3 format, you need, of course, MP3 encoding software. There are a large number of such programs available, some better than others. This software is sometimes referred to as the *converting engine*. I won't try to describe all the different systems, as there are too many. Once a recording is in MP3 format, you can put it on the Internet (or take it from the Internet). The small size of MP3 files has been a boon to the "sharing" of music over the Internet in recent years—much to the consternation of the music companies.

This brief discussion of iPods and MP3 files catches us up to the present in terms of the music scene and its physics. There is no doubt that we will see many other innovations in the decades to come—and someone else will write a new book surveying the enlarged music landscape.

SUGGESTED READINGS

Books

Askill, John. *Physics of Musical Sounds.* New York: Van Nostrand, 1979.

Boyd, Bill. *Jazz Keyboard Basics.* Milwaukee: Hal Leonard, 1996.

Burrows, Terry. *Total Keyboard.* New York: Sterling, 2000.

Cook, Perry. *Music, Cognition, and Computerized Sound.* Cambridge: MIT Press, 1999.

Esterowitz, Michael. *How to Play from a Fakebook.* Katonah, NY: Ekay Music, 1986.

Hall, Donald. *Musical Acoustics.* Pacific Grove, CA: Wadsworth/Brookes/Cole, 2002.

Hutchins, Carleen, ed. Introduction. *The Physics of Music.* San Francisco: Freeman, 1978.

Johnson, Ian. *Measured Tones.* London: Institute of Physics Publishing, 2002.

Milstead, Ben. *Home Recording Power.* Cincinnati: Muska and Lipman, 2001.

Morgan, Joseph. *The Physical Basis of Musical Sounds.* Huntington: Krieger, 1980.

Olson, Harry. *Music, Physics and Engineering.* New York: Dover Publications, 1967.

Rigden, John. *Physics and the Sound of Music.* New York: Wiley, 1977.

Roederer, Juan. *Physics and Psychophysics of Music.* New York: Springer-Verlag, 1995.

Strong, Jeff. *Home Recording for Musicians for Dummies.* New York: Hungry Minds, 2002.

White, Harvey, and White, Donald. *Physics and Music.* Philadelphia: Saunders, 1980.

Wood, Alexander. *The Physics of Music.* London: Methuen, 1962.

Internet Sites

Answers.com. www.answers.com.

Calvert, James B. "Waves, Acoustics and Vibrations." *Dr. James B. Calvert.* http://mysite.du.edu/~jcalvert/index.htm.

"Electronics Channel." *Howstuffworks.* http://electronics.howstuffworks.com.

Elsea, Peter. *UCSC Electronic Music Studios.* http://arts.ucsc.edu/ems/music.

Furstner, Michael. "Jazz Scales Lesson." *Michael Furstner's Jazclass.* www.jazclass.aust.com/scales/scamaj.htm. 2002.

Georgia State University. "Sundberg's Singing Formant." *HyperPhysics.* http://hyperphysics.phy-astr.gsu.edu/Hbase/music/singfor.html.

Global Bass Magazine. www.globalbass.com.

Henderson, Tom. "Sound Waves and Music." *The Physics Classroom.* www.physicsclassroom.com/Class/sound.

"Music." *Wikipedia.* http://en.wikipedia.org/wiki/music.

Sundberg, Johan. "The Acoustics of the Singing Voice." *ESR Acoustics.* www.zainea.com/voices.htm. March 1977.

Themusicpage.org. http://themusicpage.org.

The Violin Site: Resources for Violinists. www.theviolinsite.com.

Winer, Ethan. "Acoustic Treatment and Design for Recording Studios and Listening Rooms." *Ethanwiner.com.* www.ethanwiner.com/acoustics.html. December 2, 2008.

INDEX